Dr Richard Kent is a general practitioner and also a member of the Full Gospel Business Men's Fellowship International. He is married to Val, and they have three daughters, Emma, Sarah and Lucy. Having frequently witnessed terrible suffering resulting from cancers, strokes, overdoses and heart attacks, he was seeking the meaning of life. He was not particularly interested in Christianity, but hearing about Nazi concentration camps he was convinced that Satan was a reality, and became a Christian.

Richard became interested in the research into after-death experiences by American consultant cardiologist, Dr Maurice Rawlings, who has written the introduction to this book. This interest was the beginning of his own research into such experiences. Richard is now an international speaker on near-death experiences. Richard's profits from this book will all be donated to various Christian charities, including Tear Fund.

Val Fotherby teaches in a large comprehensive school where she is head of history. Brought up in a Christian family, she became a Christian as a young child. She is married to David who is a national director of the Full Gospel Business Men's Fellowship International, and they have a grown-up daughter, Jane. Val wrote the story of the Full Gospel Business Men's Fellowship International in *Catching the Vision*, the British sequel to the best-selling Christian biography, *The Happiest People on Earth*.

THE FINAL FRONTIER

Incredible Stories of Near-Death Experiences

RICHARD KENT & VAL FOTHERBY

Marshall Pickering
An Imprint of HarperCollins*Publishers*

Marshall Pickering is an Imprint of
HarperCollins*Religious*
Part of HarperCollins*Publishers*
77–85 Fulham Palace Road, London W6 8JB

First published in Great Britain in 1997
by Marshall Pickering

Copyright © 1997 Richard Kent & Val Fotherby

3 5 7 9 10 8 6 4

A catalogue record for this book is
available from the British Library

0 551 030984

Printed and bound in Great Britain by
Caledonian International Book Manufacturing Ltd, Glasgow

CONTENTS

ACKNOWLEDGEMENTS

The story of Dr Petti Wagner ('The Millionairess') is based on an interview with Sid Roth, Messianic Vision. The testimony of Gerard Dunphy ('The Engineering Officer') is adapted, with kind permission, from the European *Voice* magazine number 952, which is produced by the Full Gospel Business Men's Fellowship International. The story of Dr George Ritchie ('The Army Cadet') is reproduced with his kind permission.

We would like to dedicate this book to our respective spouses, Val Kent and David Fotherby. Both have provided enormous encouragement.

We'd also like to thank the various contributors to this book, especially Dr Maurice Rawlings, and Sid Roth of Messianic Vision.

And our special thanks go to Kate Davies of HarperCollins, who has provided great help and encouragement.

Dr Richard Kent would also like to thank all the members of the Harlow Full Gospel Business Men's Fellowship International, whose support and encouragement have been invaluable.

We hope and pray this book will have the powerful impact on you that researching it did on us.

Richard Kent and Val Fotherby

INTRODUCTION

A doctor explains near-death experiences

Several years ago now, I had to resuscitate a patient who died whilst trying to reproduce by exercise the chest pains he had been suffering – that experience changed both our lives forever.

The man, a 47-year-old postman, was exercising on the hospital treadmill. We were hoping the exercise would reproduce the chest pains he said he had been getting while exercising at home. But instead of just getting the pain, the electro cardiogram (the machine that shows your heartbeat) went haywire and he dropped dead, his body thrown off the still moving treadmill. I was doing the external heart compressions while the nurses sorted out a drip and breathing bag, and the patient kept saying, 'Doctor, don't stop!' Whenever I stopped to reach for something, he would scream, 'I'm in hell again!' Most patients would say, 'Take your big hands off, you're breaking my ribs,' so I knew something was wrong.

We had to put a pacemaker down his collarbone vein right there on the floor. He was writhing and blood was spurting everywhere, I was pushing and I told him to shut up and not to bother me with his 'hell business'. I was trying to save his life and he was trying to tell me about some nefarious nightmare he was having whilst in the death throes. He then asked me something which was the ultimate insult to me as an atheist, which was, 'Doctor, pray for me.' I told him he was out of his mind, I wasn't a minister. Again he asked me to pray for him and the nurses were looking at me with anticipation, and a 'you must do it, it's a dying man's wish' look. So I did. I made up a make-believe prayer, a nonsense. I just wanted to get him off my back so I told him to say it after me. I blurted out, embarrassed, 'I believe Jesus Christ is the Son of God. Go on, say it. Please keep me out of hell. Say it! And if I live I'm on the hook, I'm yours forever.' I remember that part well because he's been 'on the hook' ever since. Each time we interrupted the heart massage to adjust the pacemaker he'd scream he was back in hell, then he would convulse, turn blue, stop breathing, and his heart would stop beating.

But soon after I said the prayer there was no more writhing, no more fighting. He was calm. The next day, still highly sceptical, I asked him to tell me about being in hell. I told him he had frightened the nurses to death and scared the hell out of me. He said, 'What hell? After that prayer you said, I remember seeing my mother when she was living, although she had died when I was three years old.' Impossible! He picked her out of a photograph album his aunt brought in next day, but he had never actually seen her. He identified her from her clothing. He had seen her in heaven. What apparently happened was that he had sublimated the hell experiences to painless parts of his memory, but after the conversion he had heaven experiences.

That 'nonsensical' prayer I prayed to humour him not only got the man converted but it got me too. We both became born again Christians.

I had specialized in 'retrieval' methods long before this experience and I would teach at medical school for the American Heart Association all over the world about how to set up retrieval practices from sudden death. Provided people know what to do and the patient has not been in a 'mangled' death,

50 per cent of those clinically dead can be brought back to life again. Teaching about cardiopulmonary resuscitation, or CPR, has been going on since 1973 and is getting better all the time. The problem is that the term 'near-death experience' (NDE) has become a bit of a waste basket for all sorts of experiences, some involving bright lights but no death at all. We are trying to limit investigations to clinical death where heart and breathing stop, and then analyse the sequence of events.

The first tissue to die, because it is most sensitive to blood flow, is the brain. There is a limit of four minutes before tissue death starts occurring in brain cells. Ten minutes without CPR normally produces brain damage. There are exceptions – we have a retrieval case in Oregon where a man was submerged in icy water for 45 minutes and successfully revived with no lasting effects.

After I started researching the subject of near-death experiences, I realized about 18 per cent of the first 200 cases which we managed to retrieve were 'hell' cases and the percentage has increased subsequently. There are now a lot more people collecting information, and the reported incidence of hell experiences is now about 36 per cent and moving towards being 50–50.

Many people experiencing hell events actually block these from their consciousness because of the horror. Sometimes when being brought back, they scream about the flames of hell. One particular experience I know of was of a movie actor in Houston who was having his aortic valve (a heart valve) repaired when he had an experience of hell. He saw flames all about him and a black figure approaching. When she beckoned him over to join her, he asked who she was and she said, 'I am from the Angel of Death.' She was in the foyer of hell and he said he would not follow her. This episode turned his whole life around.

While many people who have a near-death experience of hell see lakes of fire, others just see darkness. One of the latter was a doctor who had been watching football at college. He had been so excited by a spectacular run he dropped dead. The ambulance crew assigned to the stadium treated him, but while he was 'dead' he recalls having to choose pieces of a puzzle from a conveyor belt, under a penalty if it was wrong. There were no flames but he kept screaming, 'I am in hell!' His wife was kneeling at the bedside praying him out. Because this experience literally 'scared the hell out of him', he became a Christian.

Many people have had good experiences where they see light. There is one particular book on this subject which recounts such an experience, but as a born again Christian I have trouble with it. Although it is a well accounted story the author considers our sins and faults are superfluous and that Jesus came to show love and not to seek and to save the lost. She also thinks that we all took part in the creation, assisting God, and therefore that sin is not our true nature. In my view this goes completely against scripture.

So it seems many people have these near-death experiences and believe they are in heaven and yet are not believers in Jesus. The 'Angel of Light' that they see at the end of the tunnel when they first die and get out of the body seems to welcome them unconditionally, regardless of what they have done. Theologians, on the other hand, tell us that even Satan can appear as an angel of light and deceive many, so I ask myself, whose light did they see…?

One man who had killed two people in a parking lot was himself caught, shot three times in the chest and then had this wonderful experience of light, and he later asked me whether God was a forgetful God or whether this messenger of light was not from God in the first place. This man himself questioned the appropriateness of his experience.

On the other hand there are those who have had such an experience and it serves to confirm their faith and often becomes the greatest moment of their life. Now they know what's going to happen to them when they die. So I believe that many of these 'light' experiences represent deliberate deceptions of Satan who wants people to think that Heaven's gates are open to everyone. Some people have even made a religion out of NDEs called the 'Omega Faith'. This is a case of not testing the spirit to see which light they have encountered.

Everyone wants to know what is going to happen to them when they die and life after death is what eleven million people with NDEs have claimed. Those who have had clinical death say they experienced no pain at the moment of death – they just got out of the body. Those who have had bad experiences say they are afraid of dying. They are afraid of the hell they saw.

There was one case of a blind man who, during his experience, could see perfectly well and after the transition of death he could recall who was present, what they were doing, and even what they were wearing. But when he returned to his body, he returned to his blindness.

Others report going from this world into another world through a tunnel, or something similar, and seeing a beam of light or an angel of light. People who have had car accidents often describe how they had their lives reviewed before the car crashed. It would seem possible to have a whole day's review in one split second. They then go on to the next world where they meet people, their friends, who have already died and describe strolling arm in arm across a beautiful Garden of Eden or seeing pearly white gates or golden streets, and then encountering a barrier beyond which they cannot go. Whether it's because judgement is on the other side and 'sorting out' is on this side, I don't know, but usually at that barrier they are brought back into the world of pain, back where we are pushing on their chests and breaking their ribs, or defibrillating them with paddles. Whatever we are doing it is the world of pain, and they resent it because they didn't want to come back if it was a good experience. So that's how it is – everyone who has had an NDE has had a very similar experience. Lack of oxygen cannot reproduce this, drugs can't reproduce this, nor can a high level of carbon dioxide in the blood, or anything else.

There is also commonality with those who have had hell experiences. The sequence is very fast – some zip right into the pit. For instance the father of the New Age movement, Carl Jung, recalls how the earth fell beneath him and he was in 'the place of the damned' as he called it. He saw a ball of fire in the middle of a lake and there he met Philemon, a demon.

Of people who live to tell their tale after resuscitation, 60 per cent have no experience, so only 40 per cent have had an NDE, it seems. If the person is a born again Christian, they have their dreams realized. They see Christ on the cross, and in some way they identify this being of light as Jesus Christ, the Son of God.

Dr Maurice Rawlings

POSTSCRIPT

Dr Maurice Rawlings is a specialist in heart disease. He is the clinical assistant professor of medicine for the University of Tennessee at Chattanooga, a member of the International Committee on Cardiovascular Diseases, a past governor for the American College of Cardiology for the state of Tennessee, founder of the area's regional emergency medical services council, faculty instructor for the Advanced Cardiac Life Support programmes, and Fellow of the American College of Physicians, the College of Cardiology, and the College of Chest Physicians.

He has also been the personal doctor to Pentagon staff, including Dwight Eisenhower. He is the author of several books on near-death experiences and a contributor to many medical journals.

1 THE SCUBA DIVER

(The story of Ian McCormack, New Zealand)

My whole life was centred around sport and travel. At 24 years of age, having taken a veterinary science degree at university in New Zealand, I had just completed two years travelling around the world. Now I was living in an earthly paradise for anyone who loved surfing and scuba diving – Mauritius.

I used to go surfing and fishing with the local Creole divers and got hooked on night diving. Being used to colder climates than the locals, I only wore a thin 1mm short-sleeved wet suit whereas the locals would wear the full 3–4mm suits and be totally encased from head to foot. Four days before I was due to leave the island and go back to New Zealand for my brother's wedding, I went out night diving with the local boys. I was a bit uneasy about going because I could see an electric storm on the horizon but I let myself be persuaded.

As I dived that night, the beam of my torch light picked out a jellyfish right in front of me. I was fascinated because this one was not the usual shape but 'box-shaped'. Little did I realize as I squeezed it through my leather-gloved hand, that this *box jelly-fish*, or *sea wasp*, was the second deadliest creature known to man. Its toxin has killed over 70 Australians alone and up in the northern parts of Australia, it had killed more people than had sharks. In Darwin the sting from this fish stopped the heart of a 38-year-old man in 10 minutes.

Suddenly, I experienced what felt like a huge electric shock in my forearm, like thousands of volts of electricity. Not being able to see what had happened, I did the worst thing possible, I rubbed my arm, and rubbed in the poison from the tentacles of this fish. Before I could get out onto the reef, I was stung by three other box jellyfish. My forearm was swollen like a balloon and where the tentacles had stung there were burn-like blisters across my arm. I felt on fire as the poison began moving round my body. It hit my lymph gland, as if I had been punched, and my breathing quickly became constricted.

I knew I needed hospitalization and quickly! After I was stung a fifth time, one of the divers rowed me back to shore and dumped me on the road, which was in a desolate part of the island. Lying on my back and feeling the poison taking its effect, I heard a quiet voice saying, 'Son, if you close your eyes you will never wake again.' I had no idea who said it but being a qualified lifeguard and instructor in scuba, I knew unless I got anti-toxin quickly, I would die.

My attempts to get to hospital were fraught to say the least. I had no money and an Indian taxi driver, whom I had begged on my knees for a lift, picked me up but only took me to a hotel and dumped me in the car park thinking he was unlikely to get paid. The Chinese proprietor of the hotel also refused to take me in his car, thinking the marks on my arm were from overdosing on heroin. However, a security guard, who happened to be one of my drinking companions, rang for an ambulance.

During the journey, my life flashed before me and I thought, 'I'm going to die. This is what happens before you die, your life comes before you.' Despite being an atheist, I wondered whether there was any life after death. Then my mother's face came before me and said, 'Ian, no matter how far from God you are, if you will only but cry out to God from your heart, God will hear you and God will forgive you.'

It had been 10 years since I had spoken to my mother about God, 10 years of total denial that God existed, but yet my mother was praying for me. Later, when I returned to New Zealand, I compared notes with her. God had shown her my face and said to her, 'Your son is nearly dead, start now and pray for him.' I thank God for my praying mother who had not given up on a stubborn, rebellious son. Having travelled, prior to this, through south east Asia and seen a million gods, I thought to myself, 'pray to God – which one? But my mother's face was still there and she'd only ever prayed to a Christian God. I remembered she had taught me the Lord's Prayer and thought I could just about recall it.

What followed was what happens to so many people in an exam room – my mind went completely blank, but I could hear my mother saying, 'From your heart son, pray from your heart.' 'God, if you're real,' came the prayer from my heart, 'and this prayer is real, help me remember the prayer my mother taught me. If there is anything soft or good left in my heart, please help me to remember the Lord's Prayer.' Before my eyes the words appeared, 'forgive us our sins'. I knew this meant I had to ask God to forgive all the sins I had ever committed, but I told God that I felt like a hypocrite, praying on my deathbed, but if He could possibly forgive me then I was sincere in crying out to Him to forgive my sins. It seemed as though God had heard for another part of the prayer came up, 'Forgive those who have sinned against you'. That seemed easier, to forgive people who had used me, backstabbed me or ripped me off, as I had never been vindictive or aggressive. But as I told that to God, about a foot away from my face appeared the face of the Indian taxi driver who had pushed me out of his cab that night. Could I forgive him for leaving me for dead, the voice asked? I couldn't believe it, I most certainly wasn't planning on that. I might have been planning something, but certainly not to forgive him!

Before I could think any more about him, up came the face of the Chinese hotel owner who wouldn't give me a lift either, and the voice asked if I would forgive him. What?! I realized this wasn't some head trip, this was where the rubber meets the road. I had wanted something real – well, now I'd got it and the faces of those men wouldn't go away unless I forgave them. I also realized that they were only the last two, so what about all the others before them? Knowing it was for real I promised God that if He would forgive my sins then I would forgive these men and would never lay a hand upon them. As I forgave them, their faces disappeared.

'Thy will be done on earth as it is in heaven'. I thought, 'Your will? God's will?' I'd been doing my own thing for 24 years! I promised God however, that if I came through this experience alive I would find out what His will for me was and follow Him all the days of my life. As I prayed that prayer I knew I had made peace with God and almost immediately the ambulance doors opened, I was lifted into a wheelchair and raced into the hospital.

Doctors and nurses rushed in. They attempted to take my blood pressure twice, but they could find no pulse. The doctors gave me injections of anti-toxin and dextrose in an attempt to save my life.

I was conscious of the fact that if I drifted out of my body that would be it – death. I knew this was no weird trip or hallucination, this was real, but I had no intention of leaving my body and dying. I intended to stay awake all night if necessary, and fight the poison in my system.

Feeling myself being lifted onto a recovery bed I was aware that I could not feel my arms at all and I could no longer keep my eyes open. I couldn't tilt my head, my eyes were filling up with perspiration so that I could hardly see and I remember closing my eyes and breathing a sigh of relief. At that point, from what I can ascertain from the hospital, I was clinically dead for 15 minutes.

The most scary thing for me was that the moment my eyes closed I was suddenly wide awake again, standing by what I thought was my bed, in pitch black darkness, wondering why the doctors had turned out the lights. I decided to switch the lights on and put my hand out to find the wall but I couldn't find a wall. 'OK,' I thought, 'maybe they've moved me to the general ward.' If I could get back to my bed, I could turn the lamp on but I couldn't find my bed. I thought I'd better just stand still for a moment, but it was so dark I couldn't even see my hand in front of my face and if I lifted my right hand up to my face it seemed either to miss it or go straight through. 'You can't miss your head,' I thought to myself, so I put both hands up to my face and they seemed to pass straight through. That was the most weird feeling but what followed was even worse because I realized I could not touch any part of my physical form. Yet I had the sensation of being a complete human being with all my faculties, only I didn't have a fleshly form. I realize now that I was in fact outside my body because when someone dies, their spirit leaves their body.

My next thought was, 'Where on earth am I?', because I could feel the most intense evil pervading the darkness all around me. It was as if the darkness took on a spiritual dimension. There was a totally evil presence there which started to move towards me. Although I still couldn't see, I sensed something looking at me out of the darkness. Then to my right came a voice which yelled, 'Shut up!' As I backed off from that voice another one from the left shouted, 'You deserve to be here!' My arms came up to protect myself and I asked, 'Where am I?', and a third voice replied, 'You're in hell. Now shut up.' Some people think hell is just a big party but I tell you it's going to be pretty hard to grab your beer down there, and pretty hard to find your face!

I stood there in that blackness long enough to put the fear of God into me for eternity. You might ask why God took me down there, but He told me later that if I hadn't prayed that deathbed prayer in the ambulance, I would have stayed in hell. Thank God for His grace which hears a sinner's prayer in the last seconds of his life. 'Though I walk through the valley of the shadow of death and deep darkness, yet shall I fear no evil for You are with me' (Psalm 23). I had acknowledged God as my Lord and Shepherd just before I died and He led me through that valley of death, but at the moment of deepest blackness a brilliant light shone upon me and drew me straight out. It was not like walking, but being translated up in a supernatural way. As I was drawn up into the light it seemed to touch my face and encase my entire body, as if it had pierced into the deepest darkness and pulled me out. Looking back I was able to see the darkness fading either side and could feel the power and presence of this light drawing me up into a circular opening far above me, like a speck of dust caught in a brilliant beam of sunlight. Almost immediately, I entered the opening and looking down the tunnel I could see the source of the light. The radiance, the power and purity that was flowing from it was awesome and as I looked, a wave of thicker, intense light broke away from the source and came down the tunnel at incredible speed as if to greet me. A wave of warmth and comfort went through my entire being and I felt the most incredible, comforting feeling I have ever experienced. About half way down the tunnel another wave of light broke off and came towards me. When it touched me I felt the most wonderful peace go right through me in exactly the same way as before. This was total peace. In my past I had sought for peace in education, in sport, in travel, in almost every avenue possible, yet it had eluded me. This, however, was a living peace that seemed to remain as this light left its deposit within me.

Previously, in the darkness I could see nothing but now, in the light, to my amazement, I saw my hand was like a spirit form, full of white radiant light, the same light that was coming from the end of the tunnel. I wanted to go and as I began to move another wave of light came and pure joy and excitement enveloped me. As we would call it in New Zealand, 'amped up to the max'. What I saw next 'blew my mind'. It looked like a white fire or a mountain of cut diamonds sparkling with the most indescribable brilliance. As I stopped at the end of this tunnel of light, to the left, right and above me, seemed totally filled with this iridescent light, reaching to the extremity of my vision, out into infinity. I wondered for a moment if there was a person in the centre of this brilliance, or whether it was just a force of good or power in the universe. Then a voice came out of the light and said, 'Ian, do you wish to return?' I couldn't work it out for the moment. 'Return where?', I thought, but as I looked back over my shoulder and saw the tunnel going back into darkness and thought of the hospital bed, I realized I didn't know where I was. The words came from me, 'I wish to return.' The voice responded, 'Ian, if you wish to return you must see in a new light.'

The moment I heard those words, 'see in a new light', words appeared before me: 'God is light and in Him there is no darkness at all' (1 John 1:5). They were words on a Christmas card given to me in South Africa but I hadn't known they were taken from the New Testament. As I saw these words in front of me I realized the light could be coming from God and if it was, then what was I doing here? They must have made a mistake because I didn't deserve to be here. 'If He knows my name, and He knows my thoughts as speech, I am transparent before Him. He can see everything I've ever done in my life. I'm getting out of here.' I started pulling back, looking for some rock to crawl under or go back down the tunnel where I thought I belonged.

As I pulled back from His presence wave after wave of pure light started flooding upon me. The first wave that touched me caused my hands and body to tingle as I felt love go into the depths of who I was, to the extent that I staggered. Then another wave came and yet another, and I thought, 'God, You can't love me, I've committed so many sins, I've cursed you, I've broken so many commandments.' Still the waves of love kept coming to me. Every statement I confessed was followed by another wave of love until I stood there weeping, as God's love washed through me again, and again. I could not believe that God could love such a filthy, unclean man, but yet as I stood in His presence the love got stronger and stronger until I felt that if only I could step into the light and see Him, I would know who God was.

I walked closer and closer, until suddenly the light opened up and I saw the bare feet of a Man with dazzling white garments around His ankles. As I looked up, it seemed as if the light emanated from the pores of His entire face. It was like brilliant jewels with light and power shooting out from every facet. In total wonder at the sight of the brilliance and purity before me, I realized this person indeed must be God. His garments appeared to be made of shimmering light itself. I walked up closer in order to see His eyes but, as I stood in front of Him, He moved away as if He didn't want that. As He moved I saw what looked like a brand new planet earth opening up before me. This new earth had green grass yet with the same light and radiance that was upon God. Through the fields a crystal clear river ran with trees on either side of its banks. There were green rolling hills, mountains and blue skies over to my right, and over to my left meadows with flowers and trees. It looked like the Garden of Eden or paradise. Every part of me was drinking this in saying, 'I belong here. I was made for this place. I've travelled the world looking for this place.' I wanted to enter in and explore but as I stepped forward to do so, God stepped in front of me. He asked, 'Ian, now that you've seen, do you wish to step in or return?'

Imagine, if you'd just got there, by the skin of your teeth as it were, through a deathbed prayer, and you could see a place where there would be no more sickness, no more death, no more suffering, no more pain, and no more wars. What would you do? Believe me, I had no plans to come back to this earth. I was going to say good-bye to this cruel world and step right in, but that instant I looked back over my shoulder and saw a clear vision of my mother looking at me. She had prayed for me every day of my life and had tried to show me the way of God. I realized that if I went into heaven right then she would think I had gone to hell, because she wouldn't know of my repentance in that ambulance and giving my life to God. I said, 'God, I can't step in, I can't be selfish, I must go back and tell my mother that what she believes in is real.'

Looking back I saw all my family and thousands of people stretching far back into the distance. I asked God who they were and He said that if I didn't return many of those people I could see would most likely never get the chance to hear about Him. My response was that I didn't love them but as I expressed that feeling God said, 'But I do, and I want them to come to know Me.'

How was I to get back? God told me to tilt my head, feel liquid running from my eye, then open it and see. I found myself with my right eye open and an Indian doctor at the end of the bed with a sharp instrument prodding my foot. As he turned and saw me, terror struck him, the blood drained from his face and you could see him thinking, 'A corpse has just opened its eye.'

Still trying to grasp what I had seen I heard the voice of God whisper, 'Son, I've just given you your life back.' My response to God was that if that was true could He please give me strength

9

to turn my head and look through my other eye, because I was fed up with looking at this doctor's terrified face. As God gave me strength to open my left eye, I saw in the doorway of the room a group of nurses and orderlies. They just stood in the doorway and stared, open-mouthed. I had been clinically dead for 15 minutes, but now I was very much alive!

I tried to move my neck and thought if I'd been dead that long I could be a quadriplegic for the rest of my life. I asked God to heal me completely and allow me to walk out of the hospital, otherwise to take me back into heaven. Over the next four hours I felt warmth and power flow through my body. The next day I walked out of the hospital, completely healed. I believe in healing. I believe in resurrection power. I believe Jesus Christ died for our sins on the cross, rose from the dead, and is the resurrection and the life.

What was I to do next? There was no one to ask except God and He told me I was a re-born Christian and to read His word, the Bible. Over the next six weeks I read from Genesis to Revelation and as I read through the scriptures, everything that I had seen in heaven was described in that book. In Revelation chapter 1 we read of Jesus, clothed in garments of white, His face shining like the sun, with seven stars in His hand; the Alpha and Omega, the beginning and the end. In Revelation chapter 22 I read of the River of Life with trees either side of it bearing fruit, and those who drank of it never thirsted again. I saw the light of God's presence keep the new heaven and the new earth full of light without the need of the sun or moon or lamp. This was because His radiance and presence would fill the universe. I found in John 8:12, Jesus said He was the Light of the World and those who came to Him would no longer walk in darkness but have the Light of life. As I continued to read through the gospels and the epistles, I read about being born again in John 3:3, and having the certainty of sins forgiven, of being able to call on the name of the Lord. And I knew that Jesus Christ was alive.

Since this experience in Mauritius, the Lord has led me into full-time Christian ministry. I spent some time on the dairy farm in New Zealand with my sister and her husband, and there God enabled me to get my life sorted out and I then spent six months in my local church in Hamilton. In mid-1983 I joined the evangelistic organization Youth With A Mission and for six months sailed with them through the Pacific Ocean, taking the message of Jesus Christ's love to that area. I then felt the Lord ask me to go back into south east Asia and minister to the unreached tribal people of Malaysia. For three years I worked in the jungles of Sarawak and the mainland peninsula. During this time I met my wife to be, Jane, who was on a short-term missionary trip from her home church in Canada.

Before returning to New Zealand in 1988, I worked on the pastoral staff of a church in Singapore. Jane and I were married in Canada later that year and I believe the Lord told me to take a year off from ministry and devote this time to my wife (Deuteronomy 24:5). We then worked full-time in a church in Canada. Having worked again in my local church for three years, the Lord then told me to travel and talk about my experience for three and a half years. This we have done and now hope to settle in New Zealand, serve the Lord and raise a family. We are just amazed at the blessings of God in our lives. Our desire is to continue to share the message of His unconditional love and mercy to everyone with whom we come in contact.

2 THE CAR ENTHUSIAST

(The story of Jim Sepulveda, USA)

'Jim, if you own anything of value, please make out a will,' my doctor told me after severe pains in my chest sent me to the hospital. Tests revealed an enlarged heart, a damaged main valve and two main arteries blocked by arteriosclerosis. I would need double by-pass surgery and valve replacement.

'You have a serious heart condition,' the doctor warned. I was 35 years old; I was terrified, and far too young to die!

I came from a very large, but poor family and was the youngest, and so I went around with my older brothers, one of whom worked in a car scrapyard business. When I was five years old, I went on my bicycle to this company and asked if they would give me a job. The guy laughed and told me to come back next year. I did, and he told me to come back the next year. Finally, when I was eight years old they employed me.

I wanted to be a businessman. I didn't really know what one was, but I did know kids from the other side of town who had fancy clothes and toys and big cars. They told me that their fathers were businessmen, so that's when I decided I would be one too. I worked in the scrapyard until one day, when I was about 12, I found a box of panel beating tools and I began to work on old cars, knocking out the dents. I soon became an expert and would ride my bicycle round the neighbourhood looking for cars which had dents in their wings, and then ask the

13

owner if I could fix their cars for them. A lot said no, because I was too young, but others were happy to let me have a go. I would go home on a Saturday and give my mother the money and she would get mad at me because she thought I had stolen it.

By the time I was 15, I was an expert paint sprayer. I had found an old compressor and began mixing colours and spraying the cars. It didn't seem to matter what colours I mixed, they always came out lime green! I then began my own business in my father's garage and by the time I was 16 I had pictures of my customized cars on the front covers of magazines such as *Hot Rod*, and many others. At 17, whilst still at high school, I was employing three grown men.

One day a young Jewish boy brought me his new convertible and asked me to paint it a colour that no other car had ever been, and he didn't mind how much it cost. It was so successful that I won lots of custom car shows with it, and again it was on the cover of lots of magazines. Eventually this led to my being taken into his uncle's company and working my way to the top by the time I was 32. I was then invited to join a new company and to make it profitable within four years. I did it in six months. I was still working for them, having become very successful, when I reached the age of 35 and I collapsed.

Six weeks before surgery, God began intervening in my life. I didn't know my wife, my mother-in-law and their church were praying for me! I was at home, watching television one evening, when suddenly a warm feeling came over me. I wondered if I was getting a fever as beads of sweat came from my forehead. Then, very clearly, a thought came into my mind, of a town about 35 miles from where we live. I asked my wife, Sharon, if she knew of anything going on there that night and she told me there was nothing going on there. She went out of the room but again these feelings came over me. Sharon came back three times, and each time I asked her whether she was sure there was nothing going on in the town that night. A big smile came over her face. She said there *was* something going on but it was unlikely I would want to go. Evidently there was a healing service where they prayed for the sick.

My wife had a church background but hadn't been to church for years and as for me, I had been raised a Catholic and thought people who went to healing services were a bunch of holy rollers whom I wouldn't go near for anything, not even money. In fact I laughed to myself at even the hint of any idea that I would go to such a service but even as I did so, I found myself saying to Sharon, 'Would you like to go?' I'm not sure who was the most surprised between us. Had I really suggested that?

We drove to the town and to the auditorium where the service was to be held. There was a large crowd of people going towards the front entrance so having parked the car, I insisted we sit up in the balcony, at the back, just in case there was anyone there who might recognize us. To say I felt uncomfortable would be an understatement. I kept looking at my watch, fidgeting a bit, looking at my watch again and not believing how slowly the time was going. In the end I turned to Sharon and said, 'Let's get out of here.'

What seemed to be a simple thing to do, proved impossible. As I tried to get up, a warm feeling came over me again and I couldn't move. My legs were paralysed and I thought I might be having a heart attack. Then I began to sweat, especially as the man on the stage called people forward and as he touched them they fell onto the floor. This was something weird. Again I tried to get up to go but my legs just wouldn't move. Suddenly the speaker at the front stopped and looked up. 'The Holy Spirit is telling me there is a man here who is scheduled for open heart surgery. If you will come down now, I believe the Lord is going to heal you.'

He looked around, waiting for someone to respond. He surely couldn't mean me and anyway, I still couldn't move. No one came forward and so he spoke again. 'The Holy Spirit's telling me that He wants to do something for this man. Let's pray and see if the Holy Spirit might reveal the man's name to me.' He, and almost everyone else, bowed their heads, whilst I kept looking round. After about a minute, he slowly raised his hand and pointed his finger until it seemed to be aimed right between my eyes. 'All right…Jim, come down now,' he said.

At that moment it seemed like a breath of fresh air hit me and I could move my legs. I turned to Sharon and told her I was getting out of the place and I would meet her in the parking lot. I walked out to the main aisle and up the stairs towards the exit sign at the top of the balcony. As I opened the door that warm feeling came around me again, and a very clear thought entered my mind: what have you to lose? Almost before I realized what I was doing, I'd let go of the door and was walking down the stairway towards the front of the auditorium.

The speaker asked if I believed in Jesus and I had to think. I hadn't been in church for 13 years but I said I kind of did. The next question was whether I believed Jesus had died on the cross for me and I said yes. After a few more questions, the speaker raised his arms and pointed at me. 'Jim I believe the Lord's going to heal you now.'

Yet again, that same warmth went through me, my knees buckled and I fell on the platform, but felt wrapped in a warm blanket of peace and love. Then I began to see a red light appear toward the ceiling. It came down and touched my head. A pure warm heat poured down my neck and chest, right down to my feet. An even warmer heat came up my left side and stopped in the area of my chest. Then it felt like two little fingers moved things around inside my heart and I felt physical movement inside me for about two minutes, then it stopped. Without conscious thought the words came out of my mouth, 'Jesus, I love you. I know that you've healed me. I love you.'

My doctor was not convinced when I went back to see him and told him about my experience in the auditorium. In fact he was extremely concerned and told me quite bluntly that if I didn't have open heart surgery I could die of heart disease. We discussed the situation at some length and then a thought came into my mind. 'Catheterization. Do it for the glory of God.' I knew this was a procedure during which doctors made an incision in a main artery, then fed a catheter into the heart to take pictures so they could ascertain the exact condition of the heart. I put it to him, 'Doc, listen, I don't want that open heart surgery. I want a catheterization. I want more tests.' Finally he agreed and several days later I was on the operating table.

It was one of those operations they do whilst you're conscious so I was awake the whole time and everything seemed to go well. They had just got to the last manoeuvre when I suddenly felt a searing pain in the middle of my heart. This pain ran across my shoulders and down my chest and side. As I began to lose consciousness, I could feel the doctors pounding on my chest.

'Jesus, if it's my time to come home, I'm ready,' I thought. 'I love you.' I was engulfed in complete peace with no fear of death. As a dark shadow came around me, I could hear voices from far away, echoing like a tunnel: 'We're losing him…losing him…losing him…'

I opened my eyes and I was standing in a field, surrounded by acres of green grass. Every blade glowed as if backlit by a tiny spotlight. To my right stretched a dazzling expanse of vibrant flowers, with colours I'd never seen before. Above me the endless sky was a deep and pure blue. The air around me was permeated with love.

I walked over a hill, a short distance away, then stopped beside the base of a large tree. A light began to appear beside the

17

tree. The blinding aura was too bright to look at directly. I squinted down towards the ground, and then saw a pair of sandals begin to appear at the bottom edge of the light. As my eyes moved upward, I glimpsed the hem of a seamless white gown. Higher, I could make out the form of a man's body. Around his head shone an even brighter brilliance, obscuring a direct view of his face. Even though I couldn't see clearly because of the dazzling splendour, I knew immediately the identity of this man. I was standing in the presence of Jesus Christ.

'Jim, I love you.' His voice washed over me, indescribably gentle, tender, peaceful. 'But it's not your time yet. You must go back for you have many works for Me yet to do.' I stood in awe, unable to utter a sound. Inside of me I was protesting that I was never going back, I wanted to stay right there beside Him. Almost with the hint of a chuckle, He spoke again: 'Jim, I love you but it's not your time yet.'

Then the brilliance surrounding Him reached out and engulfed me, immersing me in a total sense of love and peace. I don't know how long I stood transfixed but finally I turned away and began walking over the hill. Then a blue mist of light began to come around me like a fog. It turned into a dark shadow, and everything went black.

Suddenly I opened my eyes and realized I was lying on the operating table, covered with a sheet. I didn't know until later that I'd been clinically dead for eight minutes. Everyone had left the operating room except for the main surgeon and one of his assistants. They were at the back of the room, filling out a report on my death. After a few seconds, I sat up. The sheet slid down my lap, and I saw two men at the far side of the room with their backs to me.

'Gentlemen,' I announced, 'I'm ready to proceed if you are!' They turned and looked at me, their faces white. 'Get the rest of them in here quick,' the surgeon finally said to his assistant.

They ran test after test on me. Early the next morning, the surgeon came to my room and announced he was releasing me from the hospital. 'Come back this evening at 8.30 to my office and we'll go over all the results of your new tests.'

That evening I told my doctor what I'd experienced during those eight minutes I'd 'died' on the operating table. 'Jim,' he said after I was done, 'I'm going to show you something you won't believe.' Together we looked at the new pictures of my heart. Rather than being enlarged, it was now the normal size. Where there had been 85 per cent blockage in two arteries, there was now no arteriosclerosis and the main valve was functioning normally.

'We ran test after test on you, Jim!' He looked at me and winked. 'This is off the record…' I saw a tear form at the corner of his eye but he had a smile on his face. 'According to these pictures, this Jesus you've been talking about has either replaced or repaired your heart.'

POSTSCRIPT

Jim Sepulveda travelled the world sharing his testimony and preaching. In March 1994 God decided it was his time and at the age of 54 he died in Canada on his way home from a preaching tour.

3 THE ART PROFESSOR

(The story of Howard Storm, USA)

The date was June 1985, and I was in France. I was leading a group of students on an art tour, my wife was with me and it had come to the last day of our trip. In mid-sentence, I fell to the ground, screaming with intense pain in my stomach. An ambulance came and I was rushed to hospital to be told by the doctor that I had a hole in my duodenum, and I needed an operation. It was a Saturday. I was taken to hospital and given a bed.

With the pain getting increasingly worse, a nurse came into the room and told me and my wife that they were going to do the operation. At that point I was ready to die. I had hung on, by my finger nails as it were, trying to stay alive, but not any more.

The problem for me was that I was an atheist. As a teenager, brought up in a liberal Protestant church, I had lost faith and at college became a scientific atheist. Now, facing death, I felt nothing but hopelessness, depression and despair. I felt I was ready to die and I knew that meant I would cease to exist. I told my wife, who was not an atheist, and did have some faith, and she was in tears.

I closed my eyes and became unconscious. I don't know how much time elapsed but I found myself standing next to my body. I opened my eyes and there was a body in my bed. I didn't understand how it was possible to be outside of one's body and yet looking at that body. Not only that, but I was extremely agitated and upset because I was yelling at my wife to get her attention but she neither saw nor heard me and didn't move at all. I turned to my room-mate but got the same reaction – he too was oblivious to me and I became more and more angry and agitated. It was at that point that I heard voices calling me by name, from outside the room.

Initially I was afraid but the voices seemed friendly, and when I went to the doorway of my room I could see figures moving around in a haze; I asked them to come closer, but they wouldn't come close enough for me to see them clearly. I was able to make out only their silhouettes and general features. These beings kept asking me to come with them and although I asked a lot of specific questions, they evaded them all giving only vague answers, but insisted that I went with them. So I reluctantly agreed.

I continued to ask questions such as where were we going and they told me that I would see when we got there. I then asked who they were, but they said they had come to take me. So I followed them and we went on a journey that I know lasted many, many miles. There was no landscape or architecture, just an ever thickening, ever darkening, haze. Even though they refused to tell me where we were going, they implied they would take care of me and had something for me.

Gradually they became increasingly cruel as it began to get ever darker. The creatures also started making fun of me and some would say to others, 'Hey, be careful, don't scare him off,' or, 'Hush up, it's too soon for that.' What was even worse, they started making vulgar jokes about me. It seemed at first there were about a dozen of these creatures but later on I thought maybe forty or fifty. Still later, it appeared as if there were hundreds or more.

At this point, I said I wasn't going any further. This was a kind of bluff on my part because I didn't know which way was back, or where I was. I couldn't figure out how I could still be in the hospital but have walked as far as I had. The creatures responded by pushing and shoving me and at first I fought back well and was able to hit them in the face and kick them. However, I couldn't inflict any hurt on them and they simply laughed. Then they began with their finger nails and teeth, to pick little pieces off me. I experienced real physical pain and this went on for a long time, with me fighting and trying to fend them off. The difficulty was that I was in the centre of a huge crowd, hands and teeth all around me, and the more I screamed and struggled, the better they liked it. The noise was terrible, as was the cruel laughter and constant torment. Then they went further, insulting me and violating me in other ways too horrible to talk about and with conversation which was more gross than could ever be imagined.

Eventually I no longer had the strength or ability to fight any more and I fell to the ground. They appeared to lose interest in me. People seemed to be coming by and giving me a kick but the intense fury had gone. As I lay there I had the strangest experience. A voice that seemed to come from my chest spoke to my mind. This was an internal conversation and the voice said, 'Pray to God.'

I proceeded to argue with my voice saying I didn't believe in God so how could I pray to Him? But my voice said, 'Pray to God,' and I thought, 'But I don't know how to pray, I don't know what praying means!' For a third time my voice said, 'Pray to God!', so I thought I'd better try. I started to think things like, 'The Lord is my Shepherd; God bless America', just little things that I could remember which sounded holy. Soon the thoughts became mutterings and as they did the creatures around me started screaming and yelling at me that there was no God. They told me I was the worst of the worst. Nobody could hear me they said, so what did I think I was doing?

Because these evil creatures were so strong in their protest I started to say more, and shouted things at them like, 'God loves me. Get away from me. In the name of God, leave me alone!' They continued to scream at me except now they were retreating back into the darkness. I finally came to the point where I found myself screaming all the things I could think of that sounded religious but there was nobody around. I was completely alone in the darkness; they had retreated as if my words were scalding water on them.

Although I was shouting little pieces of Psalm 23, 'Yea, though I walk through the valley of death, I will fear no evil', and the Lord's Prayer, I didn't believe them. I meant them in the sense that I could see they were having the effect of driving these creatures off, but I wasn't convinced in my heart about the truth of them.

I was there alone. For how long, I don't know, but I sank into great hopelessness, deeper than I could imagine possible. Here I was, in the dark and somewhere out in the darkness were the evil creatures. I couldn't move, I couldn't crawl, I was too torn up and didn't know what to do. In fact I got to the point where I really did not want to exist any more.

It was at the moment of deepest despair that a tune from my childhood, when I had gone to Sunday school, started going through my head. 'Jesus loves me…Jesus loves me, this I know', and I wanted that to be true more than I have ever wanted anything in my life. With every ounce of my being, my mind, strength and heart, I screamed into the darkness, 'Please Jesus, save me!' I meant it. I didn't question or doubt it, I just meant it with every fibre of my being and upon doing that a small faint star appeared in the darkness. It grew rapidly, brighter and brighter, and soon it was a large, indescribably brilliant light which picked me up into itself. As it lifted me up I looked down at myself and saw all the rips, tears and wounds that I had received, just slowly disappear. As I continued to be lifted up, I became whole and well. The only way I can describe it is as something of inexplicable beauty which I knew was good.

One minute I was an atheist, the next minute every part of me wanted Jesus. I lost all of my pride, my egotism, my self-dependence, my reliance on my much exalted intellect. All of that had ceased to serve me any more – it had failed me. All the things I had lived my life for and made my god and had worshipped, had let me down. What I came to cry out for was a hope that was planted in a small child many years before.

I knew that the light knew me better than anyone; that it loved me in a way I'd never experienced love before and I began to cry; just completely purging myself of everything that had happened to me in my life. Until that time I had probably only cried two or three times as I considered it a show of weakness. This was the first real cry of my adult life. Now I consider crying to be very important; I give myself permission to cry if it's appropriate, I don't hold back.

This light, which I now refer to as the Angel of Light, was surrounded by other lights, angels, that came and went. Angel means a messenger from God and this was indeed the case. He held me and we rose up out of that place of darkness and started to travel through space. I saw far off in the distance what I thought was a sky full of stars but as we moved towards it I realized they were all in motion, moving towards or away from the centre.

These angels were patient, good teachers and made me feel loved and accepted, but they had some very hard things to teach me. One of the first things they wanted to do was to 'reveal' my life to me, and I told them I didn't want that because I was too ashamed of it. I had spent my life blaspheming and denying the truth, yet here I was being confronted by it. I felt the weight of all the people I had scoffed at, and perhaps as a teacher, turned away from God, denying the truth. I couldn't even bear to think of the damage I had done by my cynicism and self-worship.

Together we looked at my life, projected out in front of us in chronological order, from beginning to end. Some parts went by very rapidly, others very slowly and some parts we even watched several times from different points of view. There was no distinct background, just images of my life. It was the people who were important, not the settings. We were able to go backwards and forwards in time and see different places, but not actually be in those places.

Whenever we went to things that I had worked hard to achieve such as approval from other people, they had no interest in them and just ran them by. I would tell them to stop because I wanted them to see how hard I had worked to win that award, to see all the people watching me, but they would say, 'Yes, but that's not important.' When they came upon some incident that was bad, and there tended to be more bad than good, the angels would show it in detail. For example, one of the ways I had failed in my life was the way I interacted with people. I saw people as things to use, to get things from. In other words, I manipulated my relationships. I saw how I aggravated my father because of his interest in business. I wasn't doing it intentionally, it was teenage jealousy; I was jealous of the attention my father gave to his work and not to me. In another situation a beautiful young woman came into my life and gave herself and her love to me, but I abused her psychologically. I took for granted this love of another human being.

Another thing I saw was that God had given me the gift of children and a wife to raise them, but I saw them as extensions of my own ego. If they did what I wanted them to, if they were like me, they pleased me. If they acted in ways unlike me I would hate them and show my anger. I also saw myself constantly withdrawing more and more from people and living in my own selfish world, becoming increasingly unhappy, but getting along in the world. I was successful, getting promotions at my work, making good money and everybody thought I was a wonderful guy.

Many times the angels had to stop and simply let me know that they loved me even though I knew how much I hurt them with the life I had led, how much I had failed their expectations and hopes of what I could have been. I had seen when I was a young child that I had been made to be a loving, giving, trusting person but that I had turned away from that. It was nobody's fault but my own.

They showed me how I had turned away from the Lord. It was all pride – I didn't get good, faithful instructions when I was a teenager. What I got was a lot of extremely liberal, humanistic rationalism instead of faith. I saw myself asking people if they believed in Jesus, or in heaven and hell and they would say, 'Well, no, not really.' These were people in my church. I saw myself searching for answers and when I got to college I found people who seemed to have all the answers – Marxists and atheists. They seemed to have all the right answers about how they were going to change the world through socialism and their high-minded ideals. That was what I bought into.

There were points in my life where I could see how God had tried to reach me in so many ways. Sometimes with songs on the radio, in stories and novels I had read, in biographical sketches in history books. He tried to reach me through good people loving me and trying to open up my heart and be close to me. It seemed that every day of my life, God had reached out for me. Before this experience, if people had asked me whether God was a good God, I would have laughed at them but now I realize that God is so much better than what we think is good. Good is but a small reflection of that quality.

Having seen my whole life brought in front of me, the angels asked if I had any questions. I had, millions of them! I asked them good questions, absurd questions, intellectual questions, philosophical questions, but whatever I asked, they answered clearly and simply. People often ask how long I was with the angels and I say, 'Longer than my college education.' I know that's absurd, but that's how long it seemed. I told them I wanted to go into heaven but they said I was not ready. They said I had to go and live the way God wanted me to live. I argued as strongly as I could. They were very gentle, but adamant that for me, at that time, heaven was not an option.

I found myself back in my body and I wanted to tell my wife what had happened, but my body was so racked with pain, and I'd come from such peace and joy, that I couldn't speak to her. The nurse came in at this point and said that the doctor was going to operate on me immediately. They took my wife out of the room and I went down to the operating theatre. My wife had been so wrapped up in her sense of having lost me, because I was probably unconscious for about 35 minutes, that she couldn't really understand what was going on.

The following day when my wife came to the recovery room, I had tubes seemingly everywhere in my body. I tried to tell her about God's love and how she had to give herself to Jesus. I told her just to say 'Yes' to Him. She thought I was completely mad! When I next saw her, I tried again to tell her more calmly but I got very emotionally agitated. When the nurses came into the room I would say to them that they were doing the work of God because they helped and loved people and that God smiled upon their work. Needless to say I got the reputation of being a madman. Then I got my hands on a Bible and began to read scripture and recite it to people when they came to see me because I thought maybe my words weren't good enough, but of course people didn't like that either.

Meanwhile God spoke to me and told me to leave that hospital and come back to America. The doctors said it was a miracle I was alive; they would say the word 'miracle' and I would say, 'Do you really mean that word?' When they said yes they did, I would start telling them about God and how much He loved them and they would hurry out of the room! I had to learn, and this took many months, that my very hot zealous approach to try and convert the world was not having much success. Especially, I had to take it easy with my wife, who despite my lack of sensitivity, did come to believe in Jesus as her Saviour.

People have asked many times whether I could have dreamed all this, and there were times when I almost thought I had. When I was very sick, I would wake at maybe 3 o'clock in the morning feeling desperate and hopeless, wanting to die and be out of it all. It was then that God's love and peace would come to me and I would feel so ashamed of my negative response, and I would know that the experience I had had was given for me to believe and trust in that love. When I did that, I began to get well.

This experience changed my life completely. Not only did I eventually become a full-time minister but it changed the way I felt. Before there used to be melancholy and cynicism, but now there is genuine joy, all the time. That's not to say I don't have my ups and downs, but behind every day there is a joyfulness and I try, as best I can, to spread that joy and peace.

After a time I was invited to speak to a Bible study group and they told me that my story reinforced their faith, and I felt such love and acceptance from them that it encouraged me. From there other opportunities began to open up. I'm not important and my story is not important. What is important is that I can encourage someone in their faith, or for someone who has no faith, I can get them to re-examine who they are and what they are. It is my hope that I will be an instrument in leading them to Christ. I don't really know why God chose me to go through this experience, but as a teacher I had an ability to express things clearly and as a well-known, confirmed atheist, I think God is trying to show people His power.

I have done a lot of research in books and have interviewed people who have had near-death experiences, and have found that many are reluctant to talk about their experience because of ridicule. Also other people have gone off the deep end and made gross, wrong interpretations of what they have experienced. For example, I saw a woman on TV talking about her experience and she said, 'It's all light and love; there's no hell and no judgement, just perfect love and light.' I felt sorry for her because she had experienced perhaps a moment of the divine and had then made a very erroneous, theological conjecture from that.

What we do in this world determines where we go out of this world. People try not to face the consequences of their actions, they try to deceive themselves into saying, 'I can do whatever I want and it doesn't matter.' It does matter. Everything we do in this world matters. We can be forgiven of our wrong doing but we must be converted, which means we must renounce our sins and our guilt, and most importantly, accept Jesus Christ as our Saviour.

4 THE HILLBILLY

(The story of Rev. Ron Reagan, Tennessee, USA)

Our family was what we call a hillbilly family, living in the mountains of Tennessee. My father was an alcoholic and he abused my mother, me, and other children in the family. We had very few possessions, I was always barefoot and wore ragged clothes, but that was not unusual. When I was seven years old I mostly had to walk home from school through the mountains and at one point go through this woman's yard.

One day she came out and said she had something for me. As she led me round the corner of the house I saw a snow white lamb and it was to be mine. This little lamb became much more than a pet, it was almost my life. It would follow me and we would play together, and it would often come to the school bus stop and wait for me coming home in the evening.

This particular evening, I got off the bus and my lamb wasn't there. I ran home and as I came to the house, dad was working outside on his car, cursing, which meant he was drunk as usual. I tried to be as quiet as possible so I could slip in without him hearing me. When I walked around by the car, lying on the ground beside him I saw my little lamb, now red with blood. My dad had killed it with a tyre tool in a drunken rage. All it had wanted to do was play with him.

The hate and confusion welled up in me and I covered my ears with my hands and screamed at the top of my voice, 'He's killed my lamb!' All through the mountains rang the echo, 'He's killed my lamb!' At that moment, with the sight of that blood red coat still before my eyes, hatred for any kind of authority was born inside me.

I managed to exist, but only just, until I was 12 years old and then I ran away, working where I could washing dishes, working in roadside cafés, whatever was available. I was constantly running from the law. They wanted to take me to juvenile detention centres because I was so young, and I hated them. I would sleep in bushes and old buildings. But one night, I crawled under a big rose bush at the side of the road. It was pouring with rain and I was shivering and hungry. The black county police cars were up and down the road shining their spotlights looking for me but about a hundred yards away was a little church. I had never been into a church or introduced to Jesus. All I knew was cussing and hatred, but as I lay there I could see the lights shining and hear singing, *'Lord I'm coming home'* and *'Amazing grace how sweet the sound'*. I can remember thinking how I wished I was on the inside instead of out in the cold and wet, but I knew it was no good, I would be arrested again. Different people had tried to keep me, including my grandparents, but nobody could do anything with me because of the hatred in my life.

At 15, thinking I was a man, I stole my father's car (I had no driving licence, of course), and invited a group of similar-minded young men along. I drove towards the mountains, racing on the wrong side of the four-lane highway, crossing the double yellow lines, playing chicken, cursing, and listening to rock music. Some of us were drinking or taking dope, or both, when I remember rounding a curve on the inside with no time to stop or swerve from the oncoming car. At almost 100 miles an hour, there was a blinding head-on collision and I remember waking, lying in the middle of Chapman Highway. I looked around me and on each side I could see bodies and could hear groans, cries and screams. I was drenched in blood, it was pouring from my head. I could see the automobiles torn to pieces, yet the radio still blasted and I still hear those tunes in my head.

The Tennessee state highway patrolman came over, looked down into my face and said, 'Son, I'm charging you with manslaughter.' The following months were like a nightmare, going through the courts, hearing the screams of family members and the friends of those killed, maimed, or brain damaged for life. Inside me I knew I was at fault and that caused the hate to grow even stronger. I screamed at the judge as he sentenced me, telling him I hated him and cursing to such an extent they had to put me in chains to take me down. I wanted to bust my head against the wall. Who could have made a world like this? Who could make people this way? They sent me to the correctional institution (reform school) in Nashville.

For some reason, I managed to get out when I was 17 but was told to clear out of the county. I went up to North Carolina and got married! Elaine, my wife, was only 15, and very quickly we had two babies. My lifestyle hadn't prepared me for work and though I lied about my age to get jobs, stealing and robbery became part of my life. Elaine got involved with me and would often be the car driver – a bit like Bonnie and Clyde! My whole life was dominated by hatred and violence. It was almost as if when I was beaten, shot or cut, I really wanted to die but wasn't brave enough to take my own life. I would walk into bars and pick a fight with the biggest guy there and often lose!

When I look back at all the things my wife Elaine, and my children had to go through at that time, I find it incredible. They were afraid of me! I had many of the same attributes as my father. My wife separated from me and took the two children and was suicidal herself.

At one point we were living in Atlanta, Georgia during the sixties. It was a rough time. I rode with the biker clubs and I'd be gone for weeks at a time. Elaine wouldn't have groceries and was in terribly bad shape. I was so high on drugs that although only in my early twenties, my hair was beginning to fall out. My whole body was in a real mess through taking so many different drugs. Not only was I high on drugs but I would drink anything, often pure grain alcohol, until I was almost insane through the abuse done to my body and mind. Often I would have no idea where I was. I was about as low as it's possible to get.

Twice Elaine filed divorce papers and her family were helping her because she never knew where I was, or whether I was alive or dead. At one point she got into such deep depression that she even considered taking not only her life, but also the lives of the children. One day she took a pistol out of the drawer and was preparing to do it. All the powers of hell were telling her to kill the babies and then herself, deceiving her into thinking it was the only way out. But as she was trying to get the courage to kill the children, across the screen of our old TV, with a coat hanger for an aerial, a man called Bev Shea started singing 'How Great Thou Art', and then up stepped Dr Billy Graham to preach on the text, 'But God showed his great love for us by sending Christ to die for us while we were still sinners'. According to Elaine, there was so much power that came through those words that she put the gun away, took the children and went back to her parents' home. It was almost a year before we saw each other again.

Meanwhile I was doing everything I could to kill myself. There were so many times when I came within an inch of death. I was in at least a dozen automobile accidents at over 100 mph, drunk, high on drugs and often barely able to remember afterwards. I've been shot, knifed, and many times should have died from overdoses.

Then I became really desperate. There were a series of killings (known as the Atlanta mass slayings), and I was being accused of these. By a miracle, I escaped from the courthouse and borrowed money to ring Elaine at her parents' house. Not knowing if she would be there I was so relieved when she answered the phone. I explained the trouble I was in over the killings and that I didn't know what to do. Convinced I was losing my mind, because I couldn't think straight from one day to the next, I begged her to let me come back to her, promising I would get a job and quit the drugs. She said, 'Come home.'

We had to hide out in a little house in the mountains, and every day the police would come looking for me and I would hide, keeping away from the area until they'd gone. This went on for some weeks until finally a man confessed to the murders and I was freed from the charges. I got a job driving an 18-wheeler truck all across the country, full of liquor and dope all the time, despite all my intentions to get free.

One day I decided to take my little son, Ronnie Paul, to a town called Pigeon Ford, and a little market there. As I started to go through the entrance door to the market, another man was coming out. He wouldn't back off and neither would I. The hatred and violence just rose up in me and I busted his head right in the doorway. He fell into a stacked up case of bottles and they burst and went all over the store. People were screaming and running, but he picked up a broken bottle and came swinging for my face. As I lifted my left arm to try and stop the blow, he severed all the ligaments, tendons and the artery in my arm. In a fit of rage, I hit him again and kicked at him but this time, with that bottle, he severed the Achilles tendon and the arteries in my leg. In minutes the blood was pumping out of my body like out of a water hole. Every time my heart beat the blood would squirt out, and I quickly became faint.

The man who ran the market told me that unless I got to hospital quickly I would be dead. So he got me into the passenger side of my car while he drove, whilst my young son, watching it all, was screaming – completely hysterical. By the time we reached the hospital the floor well of the passenger side was awash with my blood – my feet were wallowing in it. I could hear voices but couldn't open my eyes any more, because all my strength had gone. When they rolled me into the emergency room I could hear the doctors and nurses saying, 'He's going to need extensive surgery. Transfer him to the hospital in Knoxville.' They loaded me into the ambulance and got me ready for transfer to Knoxville.

Someone had got hold of Elaine and she rushed to the hospital and got into the ambulance with me, as we set off. A young man, about 21 or 22, the paramedic, looked in my face and said, 'Sir, do you know Jesus Christ?' I cursed him and God, with all the strength left in my body. 'There is no God. Who is this Jesus you are talking about? Look at me. Do you think there's a God?' The young man just looked at me and said, 'He loves you. Jesus will help you. Call on Him.' Something inside of me caused me to foam and spit and cry out, 'God, if You are God, come and try me on for size.' Then something else in me would twist and I would cry, 'God, if You really exist, help me. I can't help myself. Help me please.' The young man continued saying, 'Jesus died for you, He gave His life for you.' And all the time I listened, I could hear my wife sobbing.

Smoke filled the ambulance. I couldn't breathe, I couldn't see. I thought the ambulance was on fire! 'What's wrong?' I called out, 'I can't see.' Then through the smoke I started hearing different voices, 'Razor. Razor Reagan. Ronnie! Turn around, don't come here. Go back, stop now. Don't come here!' As I kept hearing these voices, the smoke opened up and I could see what looked like the old quarry pit that we used to swim in when I was a child. In fact it looked exactly like it did on the night we poured gasoline into it and set the water on fire. It was burning and blazing and I was getting nearer and nearer to that pit. I could see people in there, and they were burning. Their arms, their faces, their bodies were blazing and the fire wasn't going out. And they were screaming my name! Closer and closer I went until I could see the individuals, but I couldn't understand what I was seeing. There were two standing close together and I saw they were Billy and Freddy, my two brothers, and they were burning and screaming. 'What are you doing here?' I yelled, 'You died on the highway in a 1957 Chevrolet, drunk, when you hit the block wall doing 100 mph. What are you doing here?' They said, 'Don't come here, there's no way out. It's horrible. Don't come here!!'

I looked to the side. 'Oh, no. Charles! Charles, what are you doing here? Last time I saw you, you were in Pigeon River. We couldn't get the car off you because we were all drunk. When you went into the river we couldn't get you out! We saw your face looking up through the water but we couldn't get you out!'

'Go back,' he said, 'don't come here.' I looked and could see flower children standing against the wall, just like I'd seen them in the sixties, dazed. Flower children so blown away. The age of Aquarius! And I saw many that had overdosed and died. Then I saw my friend, Richard. 'Oh Richard, I can't help you. When we robbed the liquor store in Atlanta, you didn't know what you were doing. You had an old pistol that didn't have any bullets in it and you didn't even ask for the money. You told the man that ran the cash register to give you a bottle of Muscatel wine. Oh, Richard! When you walked out of the door, you forgot where you were and what you were doing. But the man didn't know your gun wasn't loaded and he reached under the counter and pulled out a 357, fired point blank and blew your heart out of your chest. You fell against a parking meter and slipped down in the broken glass with the wine and the blood spilling over you. The last thing you said was, "Oh, God."' Richard cried out, 'Don't come here. You can't help.'

I cannot convey the horror, the terror of what I saw and heard. All I knew was I didn't understand it. Suddenly everything went black and I woke up.

Forty-eight hours later I came round in the hospital. My wife was sitting beside me. I had hundreds of stitches inside and outside my body. My wife explained that the doctors had decided not to amputate my arm in view of my job as a truck driver. They would keep a close watch on it though. But I wasn't interested in my arm because I remembered what I'd seen. I could not forget!

In the following weeks there was no light turned off in the house at night because I couldn't bear to be in darkness. Every time I turned a corner I was afraid I would see it all again. Me, who'd never been afraid of anything in my life. I'd been high on pain and never cared. Now I knew that I wanted to die but hadn't even the courage to do it myself. Week in and week out I tried to get stoned. I tried the booze and the dope but it didn't work.

One night I came home at 3 o'clock in the morning. I walked into the bedroom and the three children were asleep, but there was a light in the bedroom. My wife was sitting up in the middle of the bed with a big family Bible open and her face was shining brighter than the light on the ceiling. She didn't have to say anything, I knew something was different. She said, 'Honey, tonight I went to a little church with Aunt Mary and Jesus Christ saved me and came into my heart.'

'I know there's something different,' I told her, and when the next day she asked if I would go to church with her I said yes. Now I had no idea what to do. I didn't know if you had to knock on the door, I didn't know the pew from the pulpit, but I went. We sat down at the back and all the people were singing, smiling, even laughing and they were so friendly! All my life I'd sat with my back to walls – in the bar, riding in the back seat of the car if I wasn't driving. I trusted nobody. But here, something was different.

The next week I went back and the man preached as if he knew everything about me. He stood up, looked at me and said, 'Behold, the Lamb of God that takes away the sins of the world.' And I listened because he hit a note. He was talking about my lamb. 'How does he know,' I screamed inside of me. 'How dare he talk about my lamb?' As I raged inside, the preacher said, 'God will provide himself a lamb. That lamb is Jesus Christ.' I began to weep. 'Oh God, Jesus – my lamb?' 'He bled for you,' said the man, 'He shed His blood for you, no matter what you've done, no matter how bad you've been. God gave His only son, His Lamb, for you.'

By now the tears were flowing hard. I didn't want anybody to see me cry. What on earth would they think? I looked for the door and it seemed a hundred miles away. Finally that minister said, 'Come, come to Jesus and live. Old things can pass away, and all things can be made new.' I stood to my feet and started walking down the aisle towards the front, something pulling, drawing me. My heart was beating so fast and before I ever got to the front, God saved me. Now I didn't know how to pray. I didn't know fancy words. I prayed, 'God, help me or kill me! Help me or kill me! Jesus, if you're really real, help me because I can't help myself.' At that point it was as if something burst inside of me and 25 years of hatred left me. The blackness went, the demon spirits left like a black cloud and I was clean! I was forgiven!

People now ask me why I cry, run, and dance when I preach. And I think, 'Oh, Jesus, if it happened to them how it happened to me, they'd know why I'm like I am. Oh God, I don't want to hate anybody no more, I don't want to shoot anybody no more. Oh God, I love everybody.'

That was November 2 1972, at a quarter to midnight. I was 25 years old and Jesus has been real every day since. From an 8th grader in an elementary school in the Smokey Mountains, God has taken me around the world to share my story and preach the gospel. When I was in my thirties I told the devil that I was going to get back everything he had stolen from me. I learned about that in the Bible. I went back to school, finished High School, went to college and earned my BA degree. Then, just to spite the devil, I went back and earned my Masters. I want to tell you, Jesus Christ is real. He is real!

5 THE DOCTOR

(The story of Dr Dick Eby, California, USA)

My accident happened in 1972, just a week after my 60th birthday. Frankly I had never expected to achieve even this age because I had always physically overstrained myself. My wife (who has since died) and I had come from California to clear a family house following the death of an aunt. I was anxious, as ever, to get it over and done with so that I could get back to California where I had lots of appointments on my hospital and clinic books.

Coming down to the second floor from the attic with a box of debris I went through the door in the adjoining room and onto the balcony to drop the box from there as it would save me time. I leaned against the railing, which unbeknown to me had been eaten away by termites. With my weight and that of the box, the railing gave way and I plunged head first on to a cement sidewalk, two storeys down. I actually landed on the edge of the cement and the part of my head which hit the cement stopped suddenly, quite naturally. The part that was hanging over the edge continued on a bit, split the egg shell of my skull completely and broke the large blood vessel at the top of the brain. Later, I was told that some of the brain tissue was left on the sidewalk and that my body ricocheted into a bush with the heels on top. By the time the paramedics arrived I was a bloodless corpse hanging by its feet in the bush with my scalp torn loose.

As far as the paramedics were concerned, this was a corpse to take to the hospital for certification of death and then on to the funeral parlour. There was absolutely no way I could have been alive with the blood having drained from my body. Anyone can be a sceptic, but as a doctor, I'm just explaining simply what happened. It was 18 hours before I showed any evidence of life. There was no pain because there is no pain with death. Death is a sudden release from all pain, and suffering.

Instantly, but even faster than that word implies, I found myself landing with a thud in a new body, feet first, on to a solid foundation. So instantaneous was it that there is nothing we have on earth which could calibrate the speed. It seemed as though there was a feeling of vitality in this place which was so ecstatic, beautiful, loaded with love and there was peace. I knew I was in heaven because it was so fantastically different to anything we know of on earth. I was in paradise, that tiny portion of heaven where saved souls are on hold until they get their resurrection bodies. It was a place of release from all the physical difficulties that this body or mind can register and I heard myself saying, without having any ability to compose the thought, 'Dick – you're dead.'

The voice seemed to come out from me and I heard it as if I had spoken it, but it was not my voice, it was that of the one whom I knew had to be Jesus. There are so many adjectives needed to describe the qualities of this voice although in many ways it is indescribable. It was judicial, absolutely authoritative, loving, humble and kind. Conversation up there is quite different, it is mind on mind because there is no air and therefore sound would not transmit. There is nothing in the physics of heaven which is similar to the physics of matter here on earth. Down here we have five senses but in the spirit body there are so many, and you can think so fast it could not be computed. It is the same mind that Christ has.

Had you met me up there, you would have recognized me as Dick Eby because my body was of the same size and shape. The difference was that the spirit body through my spirit eyes was transparent like clear glass. When I looked to the side it would take on an opacity, but at the same time I could see right through it. It had no weight and none of the senses which register pain, fright or discomfort. There were no bones, ligaments, tissues or organs.

My mind operated fantastically differently to that here on earth. This mind is simply the product of live brain tissue and when that brain tissue suddenly dies obviously the mind is gone. When Jesus wanted to say something, I knew it immediately in my mind. If I asked a question it seemed as if He had answered it before I finished the question. We walked and talked together in paradise and just as talking is not really the right word, neither is walking. Flying would perhaps be a more adequate description because we had no weight and went simply as we wished without touching the ground.

One of the things which most excited me in heaven was the beauty of the music. As a youngster I had hoped to go into music as my profession but this was not to be. Nevertheless, my musical background helped me appreciate that the music in heaven has no similarity in sound or form to what is on earth. Up there the music flows so beautifully and has an unlimited vibration or set of waves. It is not based on a mathematical equation as is earthly music. It is of an entirely different level of hearing because we don't hear it through our ears. Instead you hear it directly in your mind. I asked Jesus who had composed this wonderful music and He asked me whether I'd read His book, because in it Jesus declares He has made everything for His pleasure but also for the pleasure of His family.

Yet another thing which amazed me was the aroma; a perfume so absolutely heavenly that it had to be made by God and for God. Later, when I was back on earth and I began reading through the Bible from Genesis to Revelation to find the answer as to what this was, I learned it was the prayers of saints. Every believer can have the great joy of knowing that when they pray their Heavenly Father enjoys those prayers so much that He changes them into a heavenly aroma. He can then enjoy this and will share it with us when we are up there.

Jesus also explained to me that He had made every individual different so that they could use their skills and abilities to fulfil their desires, and up in heaven Jesus has completed them as an individual and each individual's paradise is different. So if you are an artist, your paradise might be works of art even more wonderful than the greatest works of art on earth. If you're a musician you would want your place to reflect music and so on through all the categories which He has placed in the human concept. Jesus also explained to me that because of the new mind by which everything is instantaneous, if you want to visit someone all you have to do is to think it and you're there.

One last thing that is so different is colour. The eye of the spirit body once liberated from earthly limitations has limitless vision. This means that colours are different because the wavelengths, if that's what they are in that airless place, are limitless. All the colours of the flowers, the trees, the greenery, the sky and so forth, are so much more pure than anything on earth. But one of the more exciting things to me was when I saw that at times it can be all colours at the same time and yet you are able to see them all individually. It is so much more beautiful than anything we can perceive here.

There were so many things I wanted to know but the first was about the place I was in. The response of Jesus was in the form of a question that He was to put to me a number of times, 'My son, didn't you read My book?' He went on to explain that the Bible tells us He has prepared paradise as a holding place for saved souls until the day when God the Father would tell Jesus that the body He was preparing for Him on earth was complete. On that day Jesus said He would call all those in this area of paradise to join Him and they would descend from the third heaven, through the second heaven, through the atmosphere of the first heaven when Gabriel would sound the trumpet. Jesus said He would then call, 'Come,' and those on earth who have accepted Him as their Saviour would be taken up with Him and given a resurrection body. Everyone would come back to the third heaven, but not to paradise. This time it would be to the throne room of God.

Had God given me the privilege of staying in paradise I would certainly have taken it, but it was not yet my time and I had to return to this body. I can remember the suddenness with which Jesus, as it were, removed His mind from me and then the wonderful, brilliant light went and I was in total blackness. It was instantly cut off and later I was to find out the reason. Friends of my wife and myself were praying so fervently that I would be restored to life that God answered their prayers. There were about six prayer lines organized by one of the black churches who were the first to be told. A lovely lady from that church had been looking over the fence when I had fallen and she immediately ran in and phoned her pastor who got all these prayer lines into action.

It was about 10 o'clock on the morning of the accident that I was brought into the hospital and it was around 6 o'clock the following morning when I began to show signs of life. When I returned to my body, once more I heard the Lord speaking to me saying, 'I have come back to replace life in your body for the purpose that I put you here, and that was to continue your practice and to explain to people the things that I have shown you.' Then He started, from head to foot, putting life back into my body and we had a tremendous dialogue, at my request, because I asked Him to do it slowly so that I could see what He would do first and how He would bring life back to all my organs. This of course, was my medical training coming to the fore, because as a professor I had taught in college for many years as well as practising medicine. It was like an incredible lesson from God who had, after all, created my body. Despite my severe injuries, I have never needed any cosmetic surgery or suffered loss of memory.

Much later I learned that the hospital officials ordered that the door to my room be closed to any hospital employees. They could not write on any charts what was going on or had gone on and they didn't want their nurses and doctors to be confused because they'd never had anything like this happen before. It took five days to arrange my departure from the hospital because the airline had difficulty finding workmen to take five seats out to make room for my stretcher!

This however, was not my only miraculous experience of being outside of my natural body. Five years later I was in Israel visiting the tomb of Lazarus. This place can only hold three people at a time and I was with two women, looking at this tomb, two storeys underground. It was lit only by a small light and suddenly that light went out and we were plunged into total blackness. The two ladies screamed but I turned and said, 'Ladies, don't scream, just pray and praise.' As I said that word 'praise', something remarkable happened. Instantly the two women seemed to be taken out of the tomb and the absolute blackness was illuminated by the same light I had experienced in paradise. It was the light of heaven which originates from Christ Himself and which would burn out our optic nerves if they were in normal function. It dawned on me that the Lord must have placed me back in a spirit situation because instead of being blinded I could see perfectly.

Standing beside me was the most beautiful person I have ever seen. He was six feet tall and stood between me and the wall of the tomb. He grabbed me with His arms but then suddenly it was as if I had fallen on my hands and knees and was trying to grope for the exit. As I was doing this I noticed there were 10 toes protruding beneath the golden hem of a garment. I wanted to get up and look at this person because at the same moment (this was all happening simultaneously with great speed), I heard the voice which I had heard in heaven. I knew this had to be Jesus, so I stood up and looked into His eyes. As He looked at me it was like laser beams, such was the power of God and again, as before, He began to speak directly into my mind. He told me that He had brought me to this place to show and tell me many things.

It is impossible to describe the amazement and total surprise when Jesus then hugged me so tightly I thought my arms might break. Then I realized He was pulling me into Himself to feel the type of body He had. Now to a doctor this means something. He wanted me to realize that His resurrection body in which He was now standing was the same one that was seen by 500 people at one time, and more, as it says in the scriptures. The body had no liquid, no softness, it had no blood, serum, plasma, cerebral spinal fluid or water but was of a material which we do not know on earth. It is of course, incorruptible and eternal, and having felt thousands of human bodies, it was an amazing sensation.

The next word Jesus spoke to me shook me rigid. 'My son, I must take you to hell.' I protested and told Him that my name was written in His Book of Life. 'I know, My son,' He said, 'but I'm going to expunge it for two minutes so that I can take you and show you what the present hell is.' I told Him I didn't want to go but He said I had to because this was the end of the age and I was one who was available. Having told Jesus to get it over

47

with, I then asked if He would go with me. 'I will send My mind with you, but not My body. You'll want to ask questions and I can answer them that way,' He said.

Once again I felt myself travelling at a speed which cannot be calibrated, this time into the centre of the earth. I landed with a thud inside a hole four feet square and six feet high in solid rock. The immediate impression was one of absolute coldness, absolute blackness, and absolute silence. I immediately spoke to the mind of Jesus saying that this was different to my learning in church. His words were: 'My son, I want you to know that the present hell is a holding tank for the unsaved souls, pending the eventual judgement. God is not there. People who elect to go to hell do not want Him to interfere with their plans, they just tell Him one way or another to stay out of their lives. My Heavenly Father said He will give the desires of their hearts to all of His creatures. He grants their desires by placing them in a holding tank apart from Him until that day when they will be called and judged before Him.'

The terror was unbelievable. A spirit body cannot be frozen or burned, it simply exists with thousands of sensations which this nervous system we have on earth couldn't take. I was surrounded by demons and they seemed to be so excited they were doing a fancy rock and roll dance in honour of getting me there. Thousands of them shouted to me that I would never get out. They didn't realize I was only there for a short period and told me I was there because I had believed their deceptions whilst I was on earth. The stench was indescribable; that of dying and dead flesh. Satan, their captain, is of course the author of death. Jesus called them unclean spirits, and they certainly are.

I wanted to get out of there and screamed and yelled. As I did so, the realization came that there was no communication from the present hell to heaven. It is a place of isolation as Jesus said, and if it hadn't been for His mind being placed on mine so we could communicate, it would have been sheer terror. Suddenly however, I was snatched out of there and I found myself standing before the great white throne which St John describes in the book of Revelation. I became aware of someone surrounded by a mist whom I knew to be God, my heavenly Father. I wanted to look into His loving eyes to see the kind of love which had sacrificed His own son for me. Once again came those words from Jesus, 'Didn't you read My book? I told you in your present form you could not look upon God and live. He is too powerful and you would be extinguished. Instead you are here to be edified and educated, not exterminated. He has surrounded Himself with this mist to protect you.'

Suddenly, out of this mist, came a hand holding a book on which was written, 'The Lamb's Book of Life', and a finger started turning the pages with lightning speed. Jesus told me He was looking for my name but that He himself had expunged it for two minutes so that I could experience what it was like to be a lost sinner. Sure enough, He got to the end of the book, closed it with a clap of thunder and said, 'Your name is not in my family album. There is only one other family, that of Satan. Depart.' If I could just transmit to people the sensation that was mine when I heard God Almighty say to me, 'Depart,' then everyone would rush to make sure they knew Jesus as their Saviour.

Daniel, the Jewish prophet, tells us that when we die we do not cease to exist. Our spirit goes somewhere forever, based on our actions in this life. Once dead, it is too late to change, our fate is sealed. Daniel says we go to either everlasting life or everlasting condemnation. As we don't know when our end will come, this is the most critical decision of our life.

6 THE PATIENT

(The story of Rita Chuter, United Kingdom)

In 1969, when I was 32, I was in hospital having major surgery to my legs. The only way I can describe what happened to me is that although my body was on the table and they were fighting to get me back to life again, part of me was above them and began to float away from that operating theatre. I started to go down and down. It was horrific. I could see faces in pits, contorted with agony and pain. As I continued to go down I began to be tormented by demons of all shapes and sizes. Not only were they ugly but the smell was dreadful.

What made this downward journey even more terrifying were the lost souls I could see, and many other awful things of which I cannot bear to speak. I came into a room where I could smell sulphur and I saw a lake of fire and could feel the flames and heat from that fire. My thoughts were that if I hit the bottom I would stay for ever – never ending night and day for eternity in that dreadful place where the fear, pain and torment never stop.

It was bad enough hearing the screams and seeing indescribable horrors, but then I saw my own father in that place. I wanted so much to help but knew I couldn't do anything for him. My father had been a good man and became very sick with cancer. The vicar came to see him and asked if he could pray but my father refused any prayer or any mention of Jesus, he wanted nothing to do with Him. Now, as I looked at him, I knew he regretted that decision to turn his back on Jesus; it was his free will choice, but with what devastating consequences! As I continued to look with horror at the situation he was in, it was as if his thoughts spoke to me again, telling me to warn my mother so that she would not have to come to this terrible place where there was no respite, no end to the torment.

The heat was unbearable and I also knew there was nothing I could do to release my father from his 'chosen' destination. I was also terrified lest I should be dragged down and forced to stay there. In my desperation I cried out, 'Oh, God, please help me!'

At the moment of my crying out to God, I heard another voice saying, 'We have a heartbeat!'

I wanted to tell the surgeon and the rest of the team that I had been to hell and was terrified but I was put on oxygen and they kept telling me to rest. Nevertheless, I kept trying to tell them, 'I've been to hell, and I never want to go there again!' Yet still I did not come into a close relationship with Jesus, but allowed Satan to deceive me and to draw me away from God.

There were further major operations and each time I went into hospital I was in absolute terror of death and hell. I kept seeing myself burning in the flames, being tormented like my poor father. My mother and sister would not listen to what I had to say about dad and hell. It became a problem because I didn't go to the right people for help. I became more and more taken over by the torment and fear with which Satan and his evil spirits confused me.

It was when I was 50 years old that my lovely daughter Michelle came to see me one day and told me that yes, I was dying. My dying was not because of the physical problems but because I had allowed Satan to dominate my life. She told me that she had given her life to Jesus and that she knew of someone who would be able to help me. Thank God I agreed to allow

Canon Jim Fry to come and pray for me. On February 23 1987 my life was completely changed. As Canon Fry prayed for me to be set free from these demons, God gave me an understanding. He spoke to me saying, 'I have sent My beloved Son, Jesus, to deliver and set you free.' I saw the battle going on for my soul in the spiritual realm but Jesus dealt with it all and set me free to go and tell other people what He has done for me, to set other people free.

What a wonderful day that was. My husband was there too and he also gave his life to Jesus; we were both set free; born again by the Spirit of God, never again to be tormented by fear.

POSTSCRIPT
Rita Chuter died in 1995 and we know that she is now with Jesus.

7 THE BUSINESSMAN

(The story of David Verdegaal, Boston, England)

Life was good. I was a director in the family bulb growing and exporting business in Lincolnshire, had a lovely wife and family and a beautiful home. I travelled on the Continent on sales trips from time to time and it was in April 1986 whilst on one of these trips with my father that my life was irrevocably changed.

It was the last week of our trip to Germany and Austria and we were staying in a hotel in a beautiful town. I woke that morning feeling as if someone had stuck a knife into me and I must have gasped out in pain, because at the moment I collapsed my father woke up and immediately called for an ambulance. As the driver raced through the streets, my heart stopped beating and for 10 minutes until we arrived at the hospital, the paramedics kept me going with heart massage. The doctors at the hospital started electric shock treatment in an effort to re-start my heart. They later told me that technically, I was dead for about 30 minutes. The doctors discovered that I had a chemical imbalance in the blood and gave me a potassium injection. After another electric shock to kick-start it, my heart started to beat again.

Meanwhile my wife Jill, who was decorating the bathroom of our farmhouse, received the call to tell her that I had suffered a massive heart attack. She flew out, fully expecting to take me back with her in a coffin. I was still alive, but in a deep coma, when she got to the hospital. On the third day, according to the doctors, I suffered a stroke which further paralysed me and caused severe damage to the brain. They warned Jill that if I did survive I might suffer permanent brain damage.

For two and a half weeks I remained in a deep coma in intensive care, but at some point I went to heaven.

All my life I had gone to church and acknowledged God but I hadn't always lived the way I should and at that moment of death came the realization that I wasn't really prepared. I had always planned to get my relationship with God put to rights but now there was no more time. Even as I realized that, I also knew that I was not alone, God was there and I asked Him to forgive me for not always living my life as He would have wanted. As soon as I had prayed that prayer a tunnel of light opened up in front of me and I had the feeling of being completely changed. God not only forgave and cleansed me, but He also poured out His love on me.

I knew that I was dead but I had no fear, only a tremendous sense of security as if I had been picked up by God and cuddled in the same way as a father cuddles his baby. At this stage I could see my body in hospital but I felt detached from everything of the past. The light in this tunnel enveloped me and I just knew it was God. In the sense that we understand sight on earth, I could not see Him and yet I could see Him. He was the essence of the light.

The next thing I experienced was being led by the hand into a beautiful garden. We went through an archway of honeysuckle and saw flowers of such wonderful colour that it is impossible to describe them. This was not a large garden, it was small and compact, as if it had been prepared especially for me. Not only were the colours incredible but everything sparkled as if covered with dew and the flowers dripped light. It was so magnificent, but also peaceful, and I walked through the garden until I came to a wrought iron gate at the far end. As I came to it, the gate swung open and in an instant the whole burden of life came back upon me. I desperately wanted to go back into the garden and stay there but God told me it was not yet time for me, my life was not complete.

It took some persuading for me to return. I did not know what I was coming back to and it was in many ways a step of faith. What would happen to me? What would life be like? Whatever came, I knew that God was with me. He also told me so clearly that it was imprinted on my brain, that when I returned I must go on a retreat and I would receive further instructions for my journey.

Before I was moved from the hospital, a brain scan revealed that my brain had been damaged only in certain parts which would affect my eyesight and movement. Jill, who heard God speaking to her in the hospital chapel saying, 'He'll be healed – it'll just take time,' hung on to that hope.

At first I was paralysed, blind and dumb and it was difficult for both me and my family to come to terms with what had happened. I did not forget about going on a retreat and as soon as it was physically possible, Jill and I went. God did show me how I must face the future. He told me that I had to be willing to give up everything I held dear – my position as director in the firm, my friends, and even my family. It was not easy. I fought God, especially on the last point but eventually I said I would do as He asked. As I surrendered everything to God, He freed me to face up to the difficulties that lay ahead.

Now, eight years later, I have made a wonderful recovery. It has not been an easy time for any of us. Jill had to give up her job as company secretary to look after me and clearly there was no possibility of me returning to work in the firm. However, although registered blind and disabled, I have run two charity marathons with a sighted guide. I do have some sight and have taught myself to read again and I attend adult lecture classes to encourage my thinking and brain activity. I am also a member of St Mary's Catholic Church choir, where my family worship.

Whilst my co-ordination and memory in many areas are still poor, I have no problems in recalling my experiences during that extraordinary time in 1986.

8 THE SOLDIER

(The story of Rev. Jerry Delaney, Kentucky, USA)

As a platoon sergeant in Vietnam I was organizing the setting up of an ambush in the bean fields, about 150 yards from the jungle line, when all of a sudden a man came towards me out of the jungle. He had a weapon on one shoulder and on the other, a sack in which to collect supplies for the Vietcong. I felt an overwhelming sense of power; I could either wound him or take his life. I, and another of my platoon, opened fire and the man went down.

As an American soldier I had been trained to objectify people so that you are killing a 'thing' not a person. We were also told to take the property of the enemy when they were dead. So, in the near darkness, I went across and took the wallet out of his pocket and slipped it into mine. The following morning one of my fellow soldiers asked me what I had 'collected' the previous evening. Having forgotten about it until then, I put my hand into the pocket of my trousers and pulled out the wallet I had removed from the man – my own wallet!

About two weeks earlier, some 25 miles north east of Saigon, the 199th Light Infantry Brigade, of which I was a part, were working in the jungle villages in this area. I was leading one of several units through the jungle. The area into which we were moving was waterlogged and the water was getting deeper and deeper. Eventually I called back to one of the other officers that we needed to change direction because the water was getting too deep and we could become sitting ducks.

At this point the foliage of the jungle had become almost impenetrable so one of the men, Bill Woods, who had a machete, hacked his way through the wall of bamboo ahead of us. I followed him through but after about 10 paces he stopped, turned round to me and whispered that he thought we had walked into an ambush. The Vietcong were heading towards us in a U shape and were to the front and side of Bill and myself, the rest of the platoon having quickly moved backwards behind the thick curtain of bamboo.

By now Bill and I were up to our chins in water and suddenly he stumbled and went under the water leaving me on my own. I later discovered that he had quickly dived and swum under the water. There was a big log in the water and as I went over to the other side of it, I startled a Vietcong soldier who was hiding there. Then automatic fire started and I thought I would try and trick them into thinking I was dead. I decided to hold my breath and go into the water, making sure I came to the surface face downwards. However, when I tried to do this I discovered my feet were entangled in the roots and if that was not bad enough, I had an 80 pound rucksack on my back. At that moment the Vietcong threw two grenades into the water and it was as if someone had turned out the lights.

Suddenly, everything got really dark but I sensed that I was standing in my uniform, all neatly pressed and clean, and carrying no weapon or rucksack. Ahead of me was a long, long path and on either side there were fields and fields, as far as the eye could see, of sunflowers. The colours of the flowers, yellow, brown and green, were set against the most beautiful blue sky I had ever seen. There were no clouds, just a huge expanse of brilliant blue.

I could not understand where I was but as I looked down the path I could see, at the end of it, a small light. My whole attention was taken up by it and as I watched, it zoomed down the trail towards me. Never, in all of my life, have I seen such brightness; it is impossible to describe. It was as if the light captured me; I could not take my eyes off it and I became enveloped by it.

Gradually (though I really had no concept of time during this experience), I became aware of a presence at my right-hand side. I did not turn and look but I knew there was someone there, and somehow from within me came the thought, 'Please, don't take me, I am not ready to go.' Nothing had been said about me dying but I knew that was what it was about. Then thoughts of my mother came into my mind. I had a distant relative who had been killed in Vietnam and I knew the effect it had had upon the family. My mother could not handle my death, I thought.

As I stood there, still enveloped by this light, I heard a voice saying, 'Don't be afraid. Everything is all right.' That voice came from the right-hand side of me and, as I heard those words, I experienced a feeling of unconditional love spreading from the top of my head and down to my toes. Then came the voice again, 'He's not ready yet, you can take him back.' I felt my right hand being taken by this presence, then instantly I woke up. I was lying on the ground and all the men of my platoon were standing or kneeling around me.

The officer in charge was pressing with both hands on my chest. I began to cough and spit up water and he said in a stunned voice, 'Are you OK?' My response was, 'I think so.' He asked me what had happened – I told him what I'd experienced, but said I really didn't know. One of the other men chimed in, 'They threw grenades at you.' Another said, 'Sergeant Delaney, if you keep on like this, we'll begin to believe in that God you keep talking about.'

I told the officer I would recommend him for a Silver Star (bravery award) when we got back to base because he had risked his life to get me to safety. He looked at me totally bewildered and said, 'But I never touched you!' Thinking he was being modest, I told him he had pulled me back because I had felt his hands dragging me from the water. 'You've got it all wrong,' he said, 'I plunged into the water but I couldn't get close to you because of all the bullets, and then there was a huge explosion. By then you had disappeared and I thought we had lost you. All of sudden though, your hand came up out of the water and you were right here in front of us.' I could hardly believe it. Obviously it wasn't him, but if he hadn't come through the water to get me, then there was only one other answer – it was God's hand that had miraculously brought me through the root-filled water and back to safety.

Standing up, I fastened my clothes and we went on again and moved to a new location. Prior to this incident, I had, for some unknown reason, taken my wallet out of my trouser pocket and put it in my shirt. As I dressed I checked my shirt pocket but the wallet was gone and I knew there was no way of getting it back; it was probably lying in the bottom of the water where I had gone down.

Two weeks later we were back in the same bean fields, 150 yards from the edge of the jungle.

When I opened up my wallet, I discovered the money had gone and the photographs of my family had been put in that section of my wallet. In their place were photographs of a Vietnamese family. I cannot clearly describe my feelings at that moment when for the first time it struck me that I was killing people. This man I had killed was a family man just like me. His family would now be grieving for him. The senselessness of war hit me. All this man was doing was trying to chase me out of his country. I had great difficulty in balancing the near-death experience I had only just gone through, when I had experienced such unconditional love, and then killing another human being. I did not know how I could continue in Vietnam.

About a week later, during very heavy fighting, I was shot and lost most of my right arm. They sent me home to an American hospital to recover physically, but I suffered terrible post traumatic stress, trying to make sense of it all. In fact for a few years I was convinced I was going mad. No one I knew had ever had an experience like mine and I had certainly never read about anything like it.

I had been brought up as a Christian, married in the church, even led a church for a while. After Vietnam I went back to college and got a degree in psychology and then went on to study counselling. So deep was the trauma I had been through however, that eventually my marriage broke up and I had to go to counselling for myself. The man asked me what I was troubled by and as I tried to tell him we both sat and wept. It was the beginning of my return to sanity.

It was in 1975 I read a story in the *Readers Digest* about a lady who had had a near-death experience and the relief I felt was unbelievable. I was not going mad – it really *had* happened to me!

One great thing I have learned from all of this, is that our God is compassionate and always there to forgive. We do not have to be perfect, just repentant, and God is there to pick us up and use us for His glory.

POSTSCRIPT
The Rev. Jerry Delaney is a qualified psychologist and works in a practice but is also part of a church pastoral team.

9 THE MILLIONAIRESS

(The story of Dr Petti Wagner, USA)

I was from the cleanest family in the country – my family *were* Palmolive soap – and life had never been better. I owned property in prime areas of Houston, Texas, had a research clinic with five doctors working for me, achieved three doctorates, had four multi-million dollar corporations which I had built up myself employing over 5000 people, and was worth many, many millions of dollars.

A bizarre series of events, organized by a group of people, some of whom worked for me, led to my kidnap and murder. The intention was for this group to take over my businesses and property, once they had disposed of my body in the local bay.

I was lured to a hospital, where I thought my aunt had been taken, only to discover that she was not there. In fact, the hospital was being closed down. There were no patients in the place but I was taken by two men in white hospital coats and locked in a room on the second floor. There I was regularly beaten and my hands and fingers were terribly swollen where they forcibly removed all but one of my rings.

Their intention was to electrocute me but their first attempt ended in failure because of a power cut! Some days later, having kept me in what can only be described as primitive conditions, they tried again. This time when they hooked me up to the electroshock machine and turned on the switch, it worked. I have the death certificate to prove it.

STATE OF TEXAS			CERTIFICATE OF DEATH		STATE FILE NO.	

The certificate of death shows the following entries: Name of deceased CLIVE PEET JARVER, Sex F, Date of death MARCH 18, 1971, Race WHITE, Date of birth Oct. 30, 1915, Age 55, Social Security No 315-24-4706, County HARRIS, Street address 2223 DORRINGTON, Immediate cause CARDIAC ARREST interval March 8-18, Date signed MARCH 18, 1971, Hour of death 4 a.m., Pronounced dead on MARCH 18, 1971 at 4:20, Burial cremation removal date MARCH 18 1971.

It was a case of 'out of the body and present with the Lord'. Immediately the current was switched on I was dead. In one second I was walking across the top of the universe with my long purple velvet robe on, not yet realizing I was going to meet royalty. There in front of me was Jesus and two beautiful chairs which appeared out of nowhere, and we sat down.

Sitting there with Jesus, we were able to look down and see my body, whilst they took off the electrodes and returned it to the room in which I had been kept prisoner. We could see them coming and checking my vital signs but there were none. Jesus asked me whether I wanted to stay up there or whether I wanted to go back to earth. My response to him was, 'Lord, my job is not done.'

It may seem strange that I wanted to come back, especially when I looked at that mutilated body, as it was not something I really wanted to return to. Jesus then said, 'Any time when the blueprint of life that your heavenly Father has given you is interrupted, you have a choice. Today you are the judge, not the jury.' Immediately He said this, I was back in my body again with the sheet over my head. I was still battered and bruised, and later discovered that my hair, always jet black, had turned white due to the electric shock.

God miraculously enabled me to escape from that place and completely healed me from all the physical abuse I had received at the hands of my kidnappers. He also enabled me to overcome all the financial problems that resulted from the illegal disposal of my business assets and wealth.

POSTSCRIPT
Petti Wagner has now gone to be with her Heavenly Father, this time for ever. One of the things she used to say was that even if only one person made it to heaven, instead of hell, as a result of her experiences, then it was worth it all.

10 THE MUSICIAN

(The story of Marvin Ford, USA)

Some people would have described me as shy but that doesn't come anywhere close to it. If I had to stand up in front of just two or three people I couldn't even stutter because I couldn't say a word. I played in a jazz band in the clubs and I always stood at the back of the group because of my fear of people. One night however, the band got me drunk – I mean *really* loaded – and to their astonishment, and mine, I became the band's entertainer. If I was drunk I could entertain and talk, I could do anything, even laugh at my own jokes! Without the drink, I was the silent, back-row guy.

Not too long after this I gave up the drink because I became a Christian and didn't want that kind of lifestyle any more. I was still the silent guy at the back of the singing groups. I became even more involved with music after marrying Olive who was a wonderful classical pianist, but I was still at the back, not wanting the limelight. In the late sixties there was a big prayer convention in our church and many of the people told Olive and me they believed we were going to be involved in a world-wide ministry preaching about Jesus. Although I didn't want to say it at the time, I thought they'd got it totally wrong!

On December 29 1971, I was even more sure they'd got things wrong as I had a heart attack! I lay in an intensive care unit all night. It seemed like almost everybody I knew came into the hospital to pray for me, but I still felt as if I had an Indian elephant sitting on my chest!

The following afternoon, Thursday, something very strange happened. Hanging on the curtains around my cubicle was a big black blob. That's the only way I know how to describe it. At first I thought it was the Angel of Death; it was like a hooded monster with a huge sneer exuding from it. I realized however it was worse than that – it was the devil himself and I didn't like him because he'd caused me a lot of trouble. That 'thing' started spreading down to the floor and back up to the ceiling, completely enveloping my room. Behind it was total, utter blackness. Then it went out on the right side of my room and when it did the tube in my vein blocked.

The nurses, hearing the alarms going off, ran into the unit, took the needle out of my arm and tried to find another vein. They succeeded and I lived through that night. On Friday, exactly the same thing happened again.

A few days later, on Saturday 1 January 1972, that dark thing came into my room again. There was another man in the unit who had been defibrillated (an electric current had been applied to his chest to restore a normal heartbeat) the previous day. That afternoon, this man's heart again went into spasm and they called the doctor. So this doctor was trying to defibrillate the other patient when all of a sudden the blackness almost filled the room and the elephant on my chest had seemingly gone and fetched the rest of the herd and they were all jumping on my chest!

At that point I couldn't take any more. I had been praying ever since I had gone into the hospital that I wouldn't make false prophets out of those men who had told me I would have a world-wide ministry, but by then I didn't care any longer. I didn't care about my places of employment (I had three jobs running concurrently), the 34 years of praying had seemingly gone down the drain. I remembered the last prayer Jesus made on the cross, 'Father, into your hands I commend My spirit.' Just at that moment, my spirit left my body.

The medical profession calls it a cardiac arrest; your heart stops and the blood no longer flows to your brain. Immediately my spirit left my body, I saw the 'City of the Great King', the one in Psalm 48:1–2. I didn't go through a long tunnel, I was looking down on the most dazzling sight imaginable. All the adjectives one could use; beautiful, splendid, picturesque, colourful, magnificent, are totally inadequate to describe this place. The book of Revelation, chapter 21, gives some idea of the size which is over half of the USA. It is some 1500 miles square and surrounded by walls of solid jasper except for the 12 openings. The walls are as high as they are long and those openings were 12 gates of solid pearl, each gate of one single pearl looked as if it must have been at least 100 miles in diameter. The walls were so thick and yet not one shadow was on the outside or the inside because the brilliance of the light was so intense that nothing could cast a shadow.

In the foundations of those walls I saw precious and semi-precious stones like rubies, sardonyx, beryl, amethyst, emeralds and many others; and they were *massive*. I saw 12 columns studded with these same stones. There were three levels in that city and seemingly millions of miles of streets; avenues of solid gold, not paved with gold but solid and yet at the same time transparent. It was the purest, cleanest and brightest gold imaginable and looked like ribbons of magnifying glasses. Everywhere, through the streets and as far as I could see, were millions of mansions.

In John chapter 14, Jesus said He was going to prepare a place for us and I thought maybe the Lord had called me to be one of His building superintendents because that was my main occupation at that time! As I looked over the streets on all three levels to try and see what materials they were using, I realized there wasn't one still under construction, they were all finished. Jesus spoke those mansions into existence and He's ready; He's done His thing and now all He is waiting for is the Church to do her part.

Another thing I saw were millions of shining, sparkling, shimmering, scintillating lights which I recognized as the spirits of departed saints, all in the presence of Jesus. I knew everybody right back to the old prophets, the apostles and disciples, and they knew me. We didn't need introductions. I saw people I knew, but in a spirit form, just like I was.

What I could see was like a family of millions, with no denominations. I joined in the songs they were singing. No one was off key or out of time, and everyone was singing and worshipping God.

Having gone through all these wonderful scenes, and their rainbows of colour, I just wanted to see God. I have heard preachers preach on Romans chapter 8 about how God is 'Abba, our Father', our daddy almost, indicating an intimate relationship, but I couldn't understand this. My natural father had deserted my mother and sisters before I was three years of age and my one ambition as a boy, was to make myself strong enough physically that if I met my father face to face, I would be able to kill him with my bare hands, such was the animosity and hatred in my heart towards him because of his desertion. Thank God I never found him because I could have done it. My mother could find no trace of my father and so received a divorce by default and married again. My stepfather turned out to be the meanest man that ever drew breath and I helped make him that way.

From the day he stepped into our family we were always fighting. I hated him even worse than I hated my natural father. On the day I stood in front of him with a loaded pistol in my hand, I gave him two choices. He could disappear out of our lives or he could stand there and be shot. He left. Little wonder then, that I didn't know how to love God the Father; I had had no father image which included love.

But being in the presence of God my Heavenly Father changed all that. The only way I can describe this was like going into a massive receptacle filled to the top with the most powerful cleansing agent. It cleansed in such a way that it made everything pure and clear, like crystal without any flaw. It was as if my spirit had been covered in scar tissue that had never been erased. Although God had forgiven me, and I had later asked my own stepfather to forgive me, the scar of hatred remained until I went into the presence of God. That cleansing agent seemed to dissolve those scars and so for the first time in my 52 years I was able to say, 'Abba, Father.' Romans 8 is now one of my favourite chapters. God has complete and total charge of my life because I trust Him.

Then I had to see Jesus to thank Him for dying for me that I could have eternal life. I know it sounds strange that we believe Jesus can be up in heaven and in the earth but that is because He is omnipresent – everywhere. Jesus was seated at the right hand of God and He was not any more brilliant than God yet was distinguishable from Him. I cannot describe His looks or clothes because it was like those brilliant arc lights which shine into the skies for two or three miles. Even if you were to take several million lights, their brightness would be nothing like the brightness of Jesus. There is absolutely nothing that can compare with it, and I found myself looking and worshipping.

Something else which struck me so forcibly was that I had Jesus' entire, undivided attention. Remember, I was the bashful, timid person, an infinitesimal speck of nothing, yet I was the focus of His love. Jesus then welcomed me into His presence and I asked Him some questions. First of all I wanted to know how it was that there were millions of people constantly demanding His attention and yet I seemingly had it – undivided. Even before I could finish asking, Jesus said, 'Don't you know there's enough of Me to go around?'

I looked back through those streets of gold, through the levels of that great city and then I saw my body in the hospital, just as if someone had taken the roof of the building and opened it up underneath me. They were doing everything in their power to resuscitate me, pounding on my body, but I couldn't feel it, or even the needle they stuck in me. Then they took the oxygen mask off my nose, the suction cup from my chest, unplugged the monitor from the wall and sat down at the desk. Someone pulled out a piece of paper and began writing. From my viewpoint above the hospital I looked closer and saw written across the top of this paper 'death certificate'. When they were about two thirds of the way through filling in the form I glanced outside of the hospital and there speeding towards it, even running the red lights at the crossroads, was my pastor, Ralph Wilkerson.

That day God had spoken to my pastor, not in an audible voice, but he felt such a compulsion to get to the hospital and pray for me that he left his study before the morning services – at great speed! He got out of the car and started running towards the hospital. 'Lord,' I thought, 'here comes trouble!' I knew that he had been involved in bringing two people back from the dead and I was the third one he was about to work on. He went up to the intensive care unit and knocked on the door. At first they refused to let him in, after all I was dead! He managed to persuade them, walked over to the bed and pulled the curtains around. The corpse on the bed was the most grotesque thing you ever saw: sort of ash grey with green and purple mixed in. It also looked as if someone had pulled a stocking over my head – a very sad looking sight. He didn't put his hands on me, just clasped them together and looked up. I heard him quoting the Bible verse about the thief coming to rob, to kill and to steal but that Jesus had come so that we could have 'abundant life'. Then he quoted from John chapter 11 where Jesus told Martha that He was the resurrection and the life. And then Dr Wilkerson quoted from the book of Romans which says that if the Spirit of God who raised Jesus from the dead dwelt in us He would give life to our mortal bodies. At that point he took hold of my cold, clammy, lifeless hand and began to rebuke death and command my spirit to return to my body. It was like being at the end of a rope with which he pulled me away from the presence of the Lord.

I asked the Lord what I should do. His response was, 'What do you want to do?' I told Jesus, I'd rather stay up in heaven. Then I thought I'd make the biggest deal that was ever made. 'Lord, I'll go back if You'll go with me.'

The Lord stopped me right there. 'Don't you know I've told you in the Bible that I'll never leave you or forsake you? That I am with you until the end of the age? Since you gave your life to Me I have never left you for one split second.' I quickly changed my 'if' to 'when'. I knew He wanted me to come back. I knew He had called me to a ministry. 'When I go back Lord, will You give me a token?' He was gracious enough to tell me that as long as I went where He asked me to and preached what He asked me to, then people would always come to know Him as their Lord and Saviour.

My spirit came back into my body with impact. It was so cold and I was shaking. Pastor Ralph started patting my face and then slapped me. He pulled the covers up around my neck, then began rubbing my limbs to get the blood circulating around my body. The poor nurse! She heard the voice of a 'corpse' coming out of the room. She knew there was no way I could be talking; I had been clinically dead for 30 minutes!

I have never been sorry I came back. God did three miracles that day. Firstly, He spoke to a man who obeyed and came to pray for me. Secondly, my spirit, which had left my body and gone to be with the Lord, came back into my body. The third miracle was that whatever had happened to my brain, I liked this new one a whole lot better.

11 THE BOOKSELLER

(The story of Christine Eastell, England)

I had an excellent job as a sales representative and my career had taken off. The only way for me was up the corporate ladder. I loved the work and became more absorbed and before long instead of 'sales representative' on my card, it was 'sales executive'. I felt very proud of what I had achieved; I had made it, but at what cost! My job became more and more demanding as I had to justify my high salary, and my brand new car, and it wasn't long before the pressures of life caught up with me. Not only did I have this exciting, but demanding job, I also had two children at home and eventually it all began to take its toll.

On December 11 1989, my life was changed completely. I was suffering from a very bad dose of flu but instead of being sensible and going to bed, I decided to go into work. Everyone I went to see told me I was crazy and shouldn't be working at all but I said I was OK. The real reason was that I had just had another career move and in this new position I wasn't entitled to any sick pay, and of course the lifestyle I had become used to necessitated high outgoings each month so that even one day's pay was too much to give up.

By the afternoon I was feeling extremely ill. I had been driving for most of the day which was wet, windy and foggy – every bad driving condition you can think of. Even to contemplate driving myself home was madness but I still tried. Then, one moment's lack of concentration, a slight wandering over the central white line, and I hit the only car on the road coming in the opposite direction. Such was the impact that it took a long time to get me out of the car but through the whole time, I can honestly say that I knew God was with me. Not once did I feel any pain; in fact I felt embarrassed that everyone around me seemed to be making such a fuss. When I looked down I just saw me as me, but to others I looked absolutely terrible. The ambulance crew were amazed how calm I was though they did go through every red light to get me to the hospital as quickly as possible.

The driver of the car I hit should have had exactly the same injuries as myself but he escaped with just a cut lip and a whiplash neck injury. We had exactly the same model car and had been travelling at the same speed, and in a head-on crash of this nature, both drivers would normally receive similar injuries. He, however, got out of his car and was able to walk without any problems. Had the other driver been injured in the same way as I had, his life could have been in ruins, and I would have been responsible.

It wasn't until sometime later that I learned how seriously injured I had been. At first I felt no pain and yet it would be easier to list what I *hadn't* broken and cut! I was in a terrible mess and was lacerated all over my body. It wasn't only my car that was a total write-off, it was me too. The time in casualty was vague and unreal, as if it was happening to someone else. Over the next few days my condition deteriorated. The broken bones were not the major problem, it was my internal injuries that were causing serious concern and on two occasions my heart actually stopped and I had to be resuscitated. It was all systems go, with everybody rushing to my aid. I had gone into septicaemic shock, a type of blood poisoning, due to the internal damage.

I was taken to the intensive care unit and hooked up to machines and the prognosis was so poor that I had to have someone beside me all the time. My three grown-up daughters were there and my parents, who lived nearby, also came. They were given a room in the hospital; the staff told them to sit with me and talk to me, but they didn't know if I could hear anything. They were resigned to the fact that I was going to die, perhaps even within a few hours. I remained in that state for seven days.

Whilst I was in this limbo state with the machines in effect keeping my body alive, my spirit left my body. Everything I experienced was very clear and I know it was not the drugs I was on. I have spoken to quite a few people about it since and they assure me that the drugs would not have produced any hallucinatory effects, though for about a year afterwards I really wasn't sure.

When my spirit began to leave my body I began to go down into a very deep pit. It is difficult to describe; it was very black and misty and yet there was no beginning or end, no sides, I just knew I was in a pit. I kept closing my eyes and hoping that when I opened them it would all be a bad dream but nothing changed. It is impossible to find words to describe the fear I felt. I was desperate to get out and when I saw what I thought was a small opening I began to claw desperately. But the more I tried to get to this opening the more distant it became. It was an impossible situation.

All around there were ordinary people and they were in such deep pain and despair. They seemed to be tormented by an enormous sense of guilt which was reflected in their faces. It is so difficult to describe the depth of despair that was present in that place. If you could put all the pain, all the hurt and despair in the world together, that was what I sensed was in that place, and I was far more aware of the strength of this than I was of the people.

In the darkness I became aware of an even more evil presence than I had already felt and there, high up above anything, was this creature I just know was Satan. I didn't know what to do because this wasn't just a dream, it was reality. In my terror I screamed out, 'I'm a Christian and I belong to Jesus, I shouldn't be here,' but Satan just looked down and laughed a hideous laugh that I will always remember, and said, 'When I tempted you, you gave in to me, you belong to me now,' and at that moment I knew it was true. What could I do? There I was in hell, with Satan and in total despair. I had thought I was a Christian, but I had not committed my life to Jesus.

At that point I thought, 'Lord, please just rescue me.' I prayed for forgiveness and I remember falling on my knees pleading with Him to forgive me as I went through everything I hadn't done for Him and then I just stayed there because I couldn't do anything else. And, praise God, Satan is a defeated enemy – Jesus heard my prayer and He lifted me up from hell into His presence. I looked up and there stood Jesus in all His glory. I did not want to leave this place, this was so different to what I had come from, and now I was in the presence of Jesus and I could feel His love and His peace. I told Him I didn't want to leave, that I was happy there but He told me, 'No, you have to go back and let others know the reality of hell but also the reality of the alternative – Jesus.'

As I looked down I saw myself once again in a hospital bed, hooked up to machines and all my family sitting around and I thought to myself, 'Do I really want to go back?' Whilst I wanted to stay with Jesus I also loved my family and knew I must return. As I began to come down as it were, and return to my body I knew without any shadow of doubt that I was going to get well and I was going to do the Lord's work. I was taken off the critical list and exactly one month later, I was discharged from hospital.

Once I was taken out of intensive care I was put into a side ward and every day lots of medical students would come and visit me, talk and smile at me. I wasn't sure what was going on because I was still very ill, and at that stage I didn't realize the impact I'd had on all these people. My knees were very badly damaged and were encased in plaster and, every now and then, I would be taken down to the orthopaedic ward to get my plaster changed and they would say, 'Oh, you're Christine! We've heard about you.' I even got a mention in one of the medical journals!

When I first came round I couldn't tell them very much about what had happened; in fact they were so concerned they sent for the hospital chaplain, because what I was trying to say about my experience sounded extremely weird to them.

In the final doctor's report, which I've read, they put my recovery down to 'her very strong will to live' because that's how they saw it, but I know that it was God's will that I should live, not mine. Since that time, I have given myself 100 per cent to God and what He wants; at times it hasn't been easy but a voice would remind me, 'I wanted you alive to do My work,' and that's what I've tried to do ever since.

12 THE OSTEOPATH

(The story of Dr Ron McCatty PhD, born in West Indies, now living in England)

My mother and father felt it very important that me and my five brothers and three sisters were brought up to know right from wrong. They were very particular about our knowing the Bible, and even more particular about our knowing Jesus Christ as our Saviour before any of us left home to face a world that was largely uncaring and disinterested about such matters.

I was very close to my mother even though I didn't like many of the things she said to me. She used to say to me that I should try to be a good citizen of this world and do the things which were good and right, but that the most important was knowing that one day I would go to heaven.

Though I hate to admit it now, in those days I wasn't at all interested in turning to Jesus. I was growing up and wanted to be like all the others around. They were free to steal things and be as bad as they liked, while I had to listen to my mother telling me that the Bible taught we should not steal, and we should respect our parents, and many other such things, which to me meant only a life of restriction. I longed to be old enough to get away.

Finally that time did come and I thought that at last I had become a real person, standing on my own two feet! Now I could live like the rest of the world and enjoy the things they appeared to be enjoying. No longer did I have to listen to my mother quoting the Bible at me every day – I could put all that to one side and be like everyone else. But God had different ideas for me, and now I am grateful for the way my parents brought me up and the things they taught me.

In a way I enjoyed my life after leaving home, but I was never totally comfortable inside. I could always hear the words my mother said when I left home. 'Son, your mother's prayers shall follow you.' I was never sure whether that was a threat or a promise! I used to walk down a road and actually feel those prayers following me!

One day in December 1953 (having come to Britain), I was riding my motorbike towards the city centre, when I crashed head-on with a car right outside the gates of the local crematorium!

I was sprawled in the middle of the road, mangled and covered in blood and there appeared to be no hope for me. However, someone came, covered me up, called an ambulance and I was taken to the hospital. There they worked on me to no avail, and my sister, Dorith, was sent for. She came with a cousin to identify the corpse on the bed – me.

I couldn't understand what all the fuss was about. Dorith and Pearl were crying and calling me – and I was feeling fine. I was standing beside them. What were they bothering about? In that instant I found myself in heaven – at least at the Pearly Gates, as some people call them. How close we are to heaven, just another dimension – switch over and you're there. At the entrance, I met St Peter, St Paul and Moses, and what do you think they said? 'Sorry, stop, you are not ready.' Not ready? What did they mean, not ready? This was a dagger to my heart. I couldn't understand it. Mother used to tell me that she had offered me to the Lord when I was a baby and that I belonged to heaven. It was my spiritual heritage. She said that I needed to personally identify myself with Christ by asking forgiveness for my sins. So why was I forbidden entrance? What was going to happen to me now? I froze inside, thinking I'd missed it ... the very place which should have been my eternal home, and I wasn't allowed in. I knew instinctively that there was only one alternative – hell! I would now go down, down to the lowest hell. There was truly no hope.

But my mother had been informed of the accident and was praying, and she prayed strong, effective prayers. Now all she said was, 'O, God, I have no need of a dead son, bring him back to life'; and God heard. His answer was immediate. I was allowed to return and make my way back to earth and into my body.

POSTSCRIPT
Ron McCatty is a qualified osteopath.

13 THE MOTHER

(The story of Brenda Courtie, United Kingdom)

The year had started off in a promising way. We had moved from living in a flat to a bungalow, I was expecting a baby in the spring and we were working hard in the church. Mum had been to stay over Christmas, returning home in the new year, to a city in the grip of a flu epidemic. Mum became ill and within just a few weeks died, on my 28th birthday, in the same hospital in which I'd been born.

Her death seemed very unfair. But although I grieved as I sorted through her things, I had to keep looking forward; there was a baby arriving in the near future. Simeon was born in April and I took him everywhere, fitting in his feeds as discreetly as possible. By July, when Simeon was three months old, I was extremely tired, weepy, suspicious of people's motives and was wondering how I could possibly have post-natal depression so long after his birth.

Everything in my life became an irritation to me, or even worse, a serious problem. I had the excuse that I was still grieving but that didn't do anything to get rid of the depression. My husband John, whom I was beginning to nag unmercifully, was working hard to complete his doctorate and looking forward to possible ordination. He was very faithful at keeping our morning and evening prayer times going, but I sank deeper and deeper into depression and with that came a declining interest in spiritual things. I would nod unenthusiastically when he commented on our daily readings and would examine the hairs on the back of his hands while he prayed!

There was no point throughout this period when I thought there was no God or that Christianity was a sham. It just all seemed so distant and irrelevant. I felt trapped and I wanted to be left alone. Added to this, I began suffering with terrible digestion problems and as I lost interest in food my weight went down. Thankfully John and our friends persuaded me to seek help and the surgeon at the hospital diagnosed a long-standing diseased gall-bladder which would require surgery.

As is nearly always the case, it could not have come at a worse time. John had just started a full-time job at the college, having completed his doctorate, and I had three children to care for. Thank God for wonderful friends who stepped in and organized everything before I'd hardly had time to think about it. All I had to do was have the operation and convalesce.

'You won't know a thing,' the nurses told me as they wheeled me down to theatre. 'You'll just fall asleep, and next minute you'll wake up back in the ward.' When I succumbed to the anaesthetist's pin-prick I was transported to a world so different from the one I'd left that there wasn't even the faintest memory to remind me of it.

Despite its being new to me, I felt completely at home in this world of yellow sunlight filled with floating gold dust. It seemed as if I'd been there forever and I wanted it to stay that way. Not that this was a passive state, I was filled with hope and excitement as I could see the distant source of the light that surrounded me. I knew that the source of this magnetically intense light, flashing and bouncing off the thickening gold dust, was somewhere ahead, and all I wanted was to be drawn into it. This was of course Jesus, and I wanted to be with Him.

As I was moving towards the centre of the light something broke into this wonderful journey, something that was happening somewhere below me. Like a note of alarm jarring my peace, I heard: 'There! How's that? Any sign?' Someone else, 'No, nothing! Still nothing! O my God! Come on. *Come on!*'

Whilst this exchange was going on below me, what had started out as irritation at the intrusion, now became curiosity. Then, slowly, panic rose up in me as I realized what was going on. The surgeon and the anaesthetist wanted me to breathe. I knew that if I breathed for them, I would have to leave this wonderful light, this peace and hope, somewhere that was more 'home' than anything I'd ever known. Worse, it would mean I should have to rejoin some awful forgotten place where I had been unhappy.

Confusion gave way to compassion as I heard the desperate whisper of one of the voices, 'Oh, please, please! For God's sake, come on!' Knowing what I was doing, and what it was costing me, I took that deep breath. I drew it somewhere near the ceiling but I expelled it from the table.

The shouts of relief from those carrying out the resuscitation hit my ears from all sides. I disappeared into blackness and woke in the recovery ward.

It was difficult to try and explain to people what had happened. Fear of ridicule meant I kept quiet for some time. But I know that I had been allowed to return to my family. Had I succumbed and stayed within the light then I would never have heard the panic, have never decided to respond.

I don't pretend to understand why this should happen to me, all I know is that it did. I acknowledge that it must be part of God's will for me and my family. One thing I was certain about was that God wouldn't want me to withdraw from the warmth of His eternal light just to sit in the darkness I had been in prior to the operation. Suddenly, I could feel the September sun; I was on the road to recovery.

Throughout the long months prior to this experience I had been disinterested in prayer or Bible-reading and even now I found it difficult to get back to that close personal relationship I once had with God. One evening however, whilst on my own, I started reading through the Psalms. By the time I got into the 'hundreds' they were no longer just words. I was playing the 'Sanctus' from Verdi's *Requiem* and never was there a more appropriate accompaniment to the true joy and adoration that filled my heart as I wanted to tell God again and again how much I loved Him.

14 THE SWIMMER

(The story of Harry Newbery, Norwich, England)

I was angry with God. Very angry. I had considered myself to be a Christian; after all I went to church and took communion once a month, what more did God want? But just before I was 20, my father died. There was no way I could understand it. My father did so many good things, he worked a lot with young people like the scouts. How could God be good, as the Bible said He was, when He had taken my father from me at such an early age? From that moment it was as if I built a high brick wall and made sure God stayed behind it.

Although I now doubted everything to do with God, I still used to talk about such things. I met and married Liz who was involved with the church youth group, and we would regularly visit friends who went to the church.

April 17 1993 was a Saturday and I was preparing to take part in a scouts swimming gala, something I looked forward to. I felt great, and couldn't wait for my race to come round. About three quarters of the way down the 100 metre pool I started losing the use of my left side and had to turn on my back to complete the length. The ambulance came and the pain in my chest was like a 40 stone man pushing down on me. I kept shouting, 'Please get off me.' But whoever this 40 stone man was, he just kept on pushing and pushing down harder on my chest. I was taken to the coronary care unit, where it was diagnosed that I had sustained a massive heart attack, and for two and a half minutes I was clinically dead.

In that time, I left my body and I could see myself lying on the bed but I was surrounded by darkness. In that darkness I began to see people, milling around, and they had hoods on their heads. Suddenly a light appeared, they began to follow it and I knew I had to go with them. When I got to what I thought was the centre of the light, I saw another one, even brighter. This was the brightest, white light imaginable and in front of me, beneath the light was a pair of feet.

As I went towards the figure in the light, knowing it was Jesus, He began speaking to me and for a moment He moved to one side. At that moment I had a glimpse of a wonderful scene. There were rolling hills and countryside and flowing through this countryside was bright, running water. Not being satisfied with having glimpsed what I know to be heaven, I wanted to see the face of Jesus. He told me that if I looked upon Him I would not be able to return to my body. That didn't worry me, but He said there was a reason for me to return to this earth and I was to turn around and see. What I saw were five people, and Jesus told me I had to come back to work for Him and part of that work concerned these people.

Having made the decision to return, the next thing was I saw a brief glimpse of a few people standing around my body, one person having electric paddles in his hands. 'We've got him,' I heard them cry. Then I remember coming round, and looking at my wife and mother standing at the end of my bed.

Although I had had such a clear experience, I still doubted. Was what I had experienced truth or fiction? I had never read any books or heard stories of people to whom such things had happened. So confused was I that I found it almost impossible to tell anyone about my experience. That was until the second Monday of October in 1993. On that evening I knew a man called Ian McCormack was speaking at a dinner held by the Full Gospel Business Men's Fellowship International. He was going to tell how he died for 15 minutes, and I knew I had to go and listen to what he had to say. I had to find out whether what I had seen was truth or fiction. On two occasions, as he told his story, he broke down in tears and I was crying with him because I could relate to everything he said. Finally I knew beyond any doubt that what I had seen was the truth.

When Ian finished his story he announced that there was going to be a time of prayer and ministry and what amazed me even more was what followed. Unless the Lord had told him, he could not possibly have known because he said, 'There are two people here tonight who have heart problems and I want to pray for these two in particular. Would those two people raise their hands.' I thought there is no way he can be talking about me, but my arm went up seemingly of its own accord! The other person was a lady and because she was sitting nearer the front, she was the first to reach Ian for prayer. As I listened to Ian and his wife Jane, praying for her, I wondered what on earth was going on. Ian then turned to me. Jane placed her hand over my heart and Ian placed his hand over my head. It was as if I was there but not there; I could hear them praying but I did not seem to be in this world. I had this strange tingling sensation running through my body, almost as if the blood were draining away and yet it was a great feeling. I knew my body was swaying and all of a sudden I found myself flat on my back on the floor. Whilst lying there I had another sensation of warmth, beautiful warmth, spreading from my head down to my feet. As it moved through the whole of my body it seemed to lift every pain, every trouble I had, from my heart, and I had never felt better in my life.

Early in 1994 I was called in by the hospital for an angiogram (an X-ray technique by which the coronary arteries can be inspected). My cardiologist was amazed because my arteries were completely clear, the only damage being scar tissue in the heart muscle. He immediately discharged me to the care of my general practitioner.

I am no longer one who doubts God and His love – my doubts have been replaced by trust.

15 THE HOLIDAY MAKER

(The story of Betty Malz, USA)

I had to die to learn how to live. I was clinically dead for 28 minutes and it was during this death experience that I learned how to live. I was on vacation with my mother and dad, my kid brother Gary, my husband and my six year old Brenda and we were having a great time. The first night I experienced excruciating pain in my right side.

After I began to vomit and haemorrhage they rushed me to a little hospital. There I was met by a doctor who took a urine analysis, blood sample, and X-rays. I was kept under close observation, given pain-killing injections and transferred to another hospital, where I was given antibiotics.

I did not improve, and I had gone blind and was badly swollen. I was running a temperature of 103 to 105 degrees and was transferred to a third hospital. As they put me in the emergency room I had the feeling that I was coming there to die. Having opened me up in surgery, they discovered that my appendix had burst, and my internal organs had been affected. I lapsed into a coma which lasted for the greater part of 44 days. I had further surgery on two occasions, first for a bowel obstruction, and secondly for an abscess where the first incision had been made.

During this comatose period I was going through a wonderful learning process. I prayed most of the time though I could not speak or respond and was unconscious. I still prayed in my heart and I understood everything that was going on around me.

God used very ordinary people to bring about some extra-ordinary experiences in my life. I was a proud, haughty young woman whom God could not use, but through this suffering and 44 days of unconsciousness, I underwent a complete overhaul of personality. I always remember Aunt Gertrude coming to visit. She didn't know that unconscious people hear but she took me by the hand and very positively prayed for me, talked with me about the flowers at home, read my get well cards, the little notes that were on my flowers that had been sent to my room, and brought a little bit of the outside into that hospital room.

Another wonderful person who ministered to me was Uncle Jesse. He was a brakeman on the Pennsylvania railway and one day, heading home from Peoria, Illinois, he said he felt a very firm desire to come and see how I was doing. He thought he should go home and shower first but felt God was telling him to go to the hospital straightaway. Arriving there he learned that I urgently needed a blood transfusion. When they tested Uncle Jesse he had B-negative blood, which was the correct blood match for me, so I was given a blood transfusion from Uncle Jesse. By obeying what he believed to be the voice of God, Uncle Jesse saved my life.

In the days that followed I almost gave up as on two occasions they had to resuscitate me because I was not breathing, had no pulse or heartbeat, and they brought me back to life. Then pneumonia set in with high fever and my veins collapsed. At the end of all this they called my husband and my mother and father, and reported to them, while I listened, 'Betty's condition is very serious, and she may not live.' They all went home to prepare for my funeral and burial.

After they all left I felt as if I had suddenly got on a roller-coaster and was at the topmost point on the ride. When that sudden lurch came I realized that this was death. It wasn't frightening for it was just changing locations from this place to the other. I stood, realizing that in spite of surgery and 44 days of tubes into my stomach, I was well and strong. I was walking through a beautiful meadow of waving grass, the strands of which were like green velvet. As I walked in my bare feet, life, health and vigour began to come into my body. It was outdoors and like spring. What a joy to learn that heaven was not sitting on a damp cloud playing a harp and it wasn't wall to wall church! There was no need for the church here for I understood I was in the living presence of Jesus, the Son of God, and we would worship Him forever.

As I walked up a hill I became aware that I was not alone. A little behind me was a tall angel clad in a transparent garment of white. I saw his very capable masculine hands and a masculine face with a knowing look, and I realized he had always been there from the day that I had come to know Jesus. Walking along we talked with our thoughts – there was no need for conversation because just by wishing we understood each other and could go from earth to the galaxies of space and to the gates of heaven where, as we approached, I heard the most beautiful singing. As we reached the gates there were melodious and harmonious sounds of music coming over the wall. Suddenly I heard voices whose singing I cannot describe and I began to sing with them in a way I have never been able to do before or since. Standing there I understood all of the world's languages. The Bible verse 1 John 3:2 became alive to me, 'When we see Him, we shall be like Him for we shall see Him as He is.' He understands all of the languages being spoken all the time around the world and in His presence, I was like Him and I understood them.

Approaching a majestic gate of solid pearl, the angel touched it and it opened – I stepped inside, and saw and felt the light such as I cannot describe. Vivid yellow light, so dazzling and bright that I could not look into it. I began to strain to see where it was coming from and I believe I looked directly into the throne room of God, for seated on His right hand was Jesus. Trying to look to see His face, that brilliant light reflected on a golden boulevard and when it did, the light and the warmth of it went directly through me and I felt warm and I was healed. My past whirled rapidly before me, the present became very wonderful and real and the things of the future unfolded. In that 28 minute period I learned so much it would take many books to write about it all.

I began to look around and in that light I saw shafts of direct light ascending from the earth directly joining that great light in the throne room. It came to me that I was seeing the other end of prayer. I saw the direct shafts of light ascending from the earth; they were prayers of people on earth ascending to where God could answer either by the powers of almighty heaven or by angels or the armies of God. We have to realize that in the midst of troubles we need not be dismayed, for the powers of heaven are at our command when we pray. What made me realize this was that on one shaft of light I saw my father's voice being registered.

My father had been called to the hospital room, because I had died. He walked over to my body, with the equipment removed, and the sheet covering me. He said all he could think of to say was just to breathe one word of comfort for himself. He was asking God through this prayer to give him strength to break the news to my husband and to pray for my daughter, Brenda, when she heard the news that her mother had died. The only word he could speak was 'Jesus'. I saw it and I heard it. As Christians say, there is power in the name of Jesus. In that one-word prayer was a wish that I had not died. I saw it and felt it. I thought I would never want to leave that place but the power of my father's prayer, breathed in the form of a wish, 'Jesus', healed me and changed my mind. Now I desired to follow his prayer and to come back.

As I came back down the hill I had walked up, I looked through the roof of the hospital and saw down through the floors and into the room where my body was covered with a sheet. As I came closer I saw a direct shaft of the sun rays coming through the glass into my room. The sun was shining on my body and suddenly I felt as though I had got in an elevator and had hit the bottom floor. With that sudden lurch of inertia, I felt my body begin to warm and I touched the sheet. In the centre of that shaft of light, I saw ivory letters about two inches high coming towards me. I remembered what Art, a man who had visited me and prayed, had read, 'He sent His word and healed them'. When I looked closer I saw these ivory letters were the printed word of God from the Bible, John 11:25, the words of Jesus, 'I am the resurrection and the life and he that believes in Me though he were dead, yet shall he live.' I knew as that word was coming towards me it would heal me and Art's prophecy became reality. I touched the word, pushing the sheet off my face, and reaching out I grasped those letters pulling them into my arms and then sat up.

My first thought was that I wanted to call my grandmother who'd been very sick. When I called her I didn't know that my mother had already called her to tell her I had died. When I called I shocked her because she thought that if I was dead then she must have died too! Finally my father got on to her and said, 'Betty is back, we don't know what has happened but she is very much alive.' In the following moments I asked for a drink and for food and I was given some 7 Up on crushed ice. Pretty soon they brought in a tray on which were two pork chops and a full meal. I ate it all. Within two days I went home from the hospital.

My healing has been complete, with no side effects. I later had a perfect baby daughter. My healing was a great miracle, but even greater was the miracle that I returned with a different attitude and with a great love for people. Through the telling of this story many people have experienced the love of Jesus and experienced the greatest miracle of all, that of sins forgiven.

16 THE MISSIONARY

(The story of Rev. Royston Fraser, born in Canada, now living in England)

'Should we join the army?' the young men asked me. 'Come back next Sunday and I will preach about it,' was my response. Why not a direct answer you might ask? Well, this was the early 1970s and I was not in Canada or England, but in that long narrow strip of land, bordered by the Andes, with the longest seacoast of any country in the world, called Chile.

The country was in turmoil. In 1970 Salvadore Allende Gossens was elected to lead the country, the first freely elected Communist president in the western hemisphere. His programme aimed to extend state control to almost every area of the economy. The resulting breakdown in the economy led, three years later, to a military take-over, under General Augusto Pinochet Ugarte. This brought in a very repressive regime condemned by most human rights organizations.

I had come to Chile by a rather circuitous route. Having become a Christian whilst studying for a BSc in agriculture and horticulture, I then went on to theological college and, sponsored by the United Church of Canada, I went with a missionary team to China, only to be thrown out of there by the communist regime. For 12 years I worked with my team in India but was then asked to go to Jamaica. Later I was asked to go, again with my team, this time to Chile.

Despite the troubles of that land, we saw God move in wonderful ways. One day a young man came to me at the end of the Sunday service and said, 'I want to play football.' I told him that I believed when a young boy had been to church on Sunday it was a good way of keeping him out of mischief. He then said, 'What, with this?' As he lifted up his leg I saw he had a club foot. I asked if I could pray for him and when he agreed I sent him to get another leader of the church so we could pray together. Almost immediately, the young boy started screaming and I told my colleague to take off his boot. I could not pray any more because all three watched as his foot grew and became normal.

The young boy had some way to go home and I knew he couldn't walk with one boot that fitted, so I went to a nearby shoe shop and explained the situation saying I had only $11. The owner was so astounded by what he heard that he said I could have some boots for that amount. A few years ago the Chilean football team came to London and I received a letter telling me to be there and that I would receive the shock of my life. Not only was the young man playing football for his country but he also scored a goal!

Every Sunday service we would have at least one member of the secret police taking notes of what was said, and this was why I could not give a direct answer to these young men. On the Sunday morning following the question, I preached on the well-known text, 'Render to Caesar that which is Caesar's, but render to God that which is God's.' I thought I had got round the problem quite well but that was not the thinking of the authorities.

The following Sunday, we had just started our service when about 12 men burst in, guns blazing, and they shot about 25 of the congregation. My deputy, John, was one of them. Seeing them coming towards me, he jumped in front to shield me from the bullets. If ever there was a case of 'Greater love has no man than this, that he should lay down his life for a friend', then this was it. John was killed instantly but five bullets went straight through him and exploded in my stomach.

I knew I desperately needed medical help. The British consul said he could get me to England but it would have to be there and then. I left on the aircraft that afternoon with only the clothes I had on and my Bible, which I still have. By the time we reached England I was unconscious and totally unaware of them putting me in a helicopter for an emergency landing at the hospital. They were all prepared for me and I was taken straight into the operating theatre.

I became aware of where I was when quite suddenly I found myself standing in the air – suspended as it were in space – looking at my body on the operating table. After a few moments two angels came along and said, 'This way.' We travelled along a dark passage but all the time I could see a light at the end. When we got there it was like going through thick cobwebs and into the most marvellous place I have ever seen.

The first person I saw was my mother. You might think that not surprising, but it was to me. Sometime before leaving Chile I had a letter from my parents saying they were moving over to England and were going to settle in Norfolk. What I didn't know was that whilst travelling my mother suffered a massive heart attack and survived for only a week after arriving in England. She came forward and threw her arms around me and hugged me.

Another surprise awaited me as I looked across and saw my sister who waved. Again, I did not know she was dead, though I knew she had been shot in the foot in Chile. Unfortunately gangrene had set in and she died.

The two angels were still with me and started to introduce me to people, some missionaries, and some well-known figures of history. What I was aware of was the splendour of the place which was wonderful. Everything was dazzling, the colours, especially the green, were so incredible it is indescribable.

Finally, we came to the one person to whom I needed no introduction – Jesus. As I looked at this majestic figure, love seemed to pour out from Him. The only words I can find are those of a song we sing, 'Beautiful beyond description'. I asked what might seem a strange question; it was, 'Where am I, Lord?' His response was, 'You are in the paradise of God.' I told Him I thought I was in heaven but He replied, 'No man has entered heaven except He which has come out of heaven.' I said I hadn't really thought about it but I did recall reading it. 'Don't you remember what I said to the man on the cross?' He asked me. Of course I did: 'This day you will be in paradise.'

Thinking this was all wonderful, I asked, 'What happens next?' The reply of Jesus was, 'For you, nothing.' Then one of the angels spoke and said, 'Not this time.' As I heard those words I went all the way back into my body.

The following morning the surgeon came and spoke to me and asked if I knew I had been clinically dead for nine minutes. I told him that I knew I had died but was unaware of the time because the moment it happened, I was in eternity.

I recovered from my ordeal but was told I must not go back to being a missionary but should find some very light work to do. So, as always, I prayed about it and asked God to open up an opportunity for me. The very next day a man came to see me and said, 'What are you going to do now?' I told him I was going to find a job and then asked who he was. His reply was not what I expected. 'I'm a Christian and I've heard about you. At present I'm the playing field officer for Buckinghamshire County Council but as I'm leaving the job I need someone to replace me. I know you have a BSc in agriculture and horticulture and I think you're the man for the job.' I took up the post and at the same time ran a small church.

I finally gave up the job when I had a heart attack but today I am still active and am the prayer officer for a chapter of the Full Gospel Business Men's Fellowship International. God is so amazingly good!

17 THE BEREAVED MOTHER

(The story of Jeanette Mitchell-Meadows, USA)

In the space of just two months in 1977 my world turned upside down – my daughter and best friend died, and I was not only diagnosed with glaucoma, but broke my back. My back was broken when I slipped on the floor. The surgeons decided to 'fuse' it together.

Just prior to this, my daughter was hit by a truck in West Virginia and killed. Annette was a lovely Christian girl, who loved Jesus so much. We prayed for all our worth that she would live. Many other people were also praying but she did not recover. I now believe she died because she had a choice. One of the people who came to pray for her said, 'She's in heaven right now,' and I knew that was right and that she had made her choice to stay.

Now I can understand, but at that time, I didn't want her to leave us. She was our only child. We loved her but she had a free will and I know how much she loved God and she is with Him. Her daddy was devastated, pounding on the walls of the hospital till they vibrated. Annette's death, of course, made the fact I was having an operation even more difficult for him.

When I went for surgery the operation took a long time. On many occasions when I was in the recovery room, I stopped breathing over and over again. My spirit left my body during the operation and altogether I was in the presence of the Lord for nine hours.

In the time it takes to blink an eye I was in heaven and saw Jesus. There was such peace, and the light was so bright that if you were looking at Him in your normal body, you couldn't stand it because it is so glorious. I can't really judge how tall He was, but His hair was dark, sort of brownish. What was so over-whelming was the love which just permeated His whole being. This love and peace engulfed me to the extent that I did not want to come back even after I had agreed I would. I understood then how my daughter had chosen to stay.

I saw the gates of heaven, made up of 12 gigantic pearls, and the streets looking like solid gold. The walls are precious stones and all the colours are so bright and vibrant. As well as all this, there is the music which is almost impossible to describe. It is glorious; it is worship and adoration of God; it is holy and pure, and you couldn't try and sing to keep up with it in the natural way. There were musical notes I have never heard on earth, so clear and flawless and the tone was so beautiful. It was the most wonderful place to be.

After a little while I saw my daughter. We were in a garden where the grass was so green and lush, the flowers incredibly bright and fruit trees everywhere. Someone picked an apple from one of the trees and it immediately grew back. We talked about how much we loved each other and how happy she was to see me. The feelings were mutual, but of a higher level than we understand on earth.

My grandparents were also in heaven, though in a different place to my daughter. Up there, you don't just float on a cloud but everyone is an integral part of what God is doing and they all have work to do. There were others of my family there, my great-grandma for one, but there were not many because God told me that they didn't make it to heaven. There were some people I expected to see there but they were not there.

Jesus spoke to me and told me He loved me. We talked about

many different things and then He said He wanted me to come back. This did not please me and I asked why I couldn't stay. He told me He had something for me to do and when I had finished it then I could come back home again. Jesus did not say what this would be but that I would know each step of the way. It was then that I started to cry but I had given my word and agreed to come back.

My return was somewhat slower than my journey to heaven, and on the way back I experienced no pain whatsoever. When I was fully back into my body again the pain was excruciating. I couldn't think at all; I had no language; the only words I knew were 'yes' and 'OK'. I was told later that this was due to a lack of blood and oxygen to the brain.

After the four hours of surgery, six people worked on me for another five hours non-stop! Because I had started to cry when I came back into my body, the nurses wanted to know if I was upset about what they had said among themselves. They knew that a patient can hear you speaking even if apparently unconscious. I shook my head to try and indicate that I wasn't upset with them but I knew I was crying because I was alive. I guess I really didn't want to come back to earth and leave my daughter behind. When I was walking round in heaven there was no pain, I had no brain damage, my head wasn't foggy, I was alert and could communicate. Now it was very different!

After the surgery I had kidney failure; my lips peeled and bled for two months; I didn't know my alphabet. I couldn't even eat properly – I kept hitting my face with the fork. The doctors said that I had sustained brain damage which might not improve. I didn't want to be a vegetable and my husband, already angry after the death of our daughter, became even more angry with God. Seeing the state I was in hurt him so badly and he couldn't handle it.

When I first came back home my husband took me out and bought me lots and lots of presents, even things I couldn't use at that time. Gradually he realized how ill I still was and he got more distant and hurt and angry. It came to a point that he couldn't stand to see me suffer. He wouldn't even pray because after Annette died he figured there was no point. Eventually he left me.

It was a very slow process of healing from then until 1983, though I saw many miracles on the way. In 1983, I arranged to visit my sister in Florida for Thanksgiving. For the three weeks before that visit, I looked up every Bible verse on healing and the New Testament passage of Acts 3:6 in particular became very real to me. This is where Peter and John tell the beggar at the temple gate that his faith in the name of Jesus has made him whole. I meditated on those words at least a hundred times a day. Then whilst in Florida, I went to a church in Fort Myers and went out for prayer from the pastor. He came over to me and asked me what I needed. 'Well,' I said, 'I died, broke my back, have glaucoma and lots of other things!' He just took things one at a time and then asked if I had faith to believe for the healing of my back. The scriptures say, 'According to your faith be it unto you.'

When he prayed, the intense pain in my lower back left instantly. Up to then it had hurt to sit and to lie down. In fact many times I slept on the floor with my feet in a chair, but I could get no real relief. After the prayer I ran round the church! Not only was my back healed but my glaucoma too. I felt something lift from my eyes and have never used the eye drops since. I have been examined and told that the glaucoma has gone.

I believe that God sent me back so that I could minister life to people as an example of one who had received many, many miracles, and help them build up their faith in a living God.

18 THE ENGINEERING OFFICER

(The story of Gerard Dunphy, Ireland)

This was a headache of mammoth proportions! It was May 20 1982, and I was at home in bed. Pam, my wife, had gone to town and I was alone with our 15 month old baby. As the morning wore on, the headache grew even worse until suddenly I felt something snap in my head. I didn't know what it was but I knew it was serious.

Knowing I needed help, I called out to God. There were a number of things not right in my life so I promised, 'If I get through this problem, I'll straighten everything out.'

At this point I became concerned for the baby, who was on my bed. I managed to get hold of her by the leg, put her onto the floor, and drag her down the landing to the back room, where I locked her in. I then crawled back to my bedroom, got to the window, and called out to my neighbours, who were outside at the time, for help.

I was sent by ambulance to the hospital, where I was examined and had a lumbar puncture. A brain haemorrhage was diagnosed. They transferred me to the regional hospital, where the specialist arranged for a CAT scan. I had an angiogram the next morning to look for blocked arteries and woke up nine hours later in the recovery room after surgery. Then I lapsed into a coma. The prognosis was poor. I was put onto a life-support machine but as the clock ticked on I slipped into a deeper coma.

Evidently I was in that coma for days with no change except that the specialist told my wife I did not have fully fixed or dilated pupils, nor a temperature over 106 °F. Pam did not understand the medical terms so asked a cousin of hers who was a nursing sister. She said that I was on the brink of death but had not quite reached the point of death. The cousin asked Pam if she'd ever heard of the healing power of God, and slipped a piece of paper with a telephone number into her hand.

That night Pam rang the number and spoke to a man, Al Ryan, who was the president of the local chapter of the Full Gospel Business Men's Fellowship. When she told him the problem there was a moment's silence, then Al said, 'Mrs Dunphy, your husband will recover.' Somehow Pam knew, deep down, that his words were true.

Al then asked for three days of prayer and fasting. Pam's cousin contacted as many people as she could, requesting their prayer support. It was put out on the local radio's Christian programme, and prayer groups around the country were contacted. That Sunday evening all of my family gathered to pray for me. This kind of meeting was new to the family. It was a last resort! What Pam didn't know at that time was that all my family had actually gathered to say their goodbyes to me. My brother had even come over from England.

The night after the prayer meeting, Pam and her cousin came to visit. They noticed a slight change in my condition. When they spoke to me, my eyelids fluttered. Her cousin went out and spoke with the people monitoring my condition and asked if there had been any change. No, they said. Two hours after they left the hospital contacted my family. I was conscious. I knew where I was, what had happened to me and even my name! I could see clearly, and even read. The next morning the specialist informed my wife, 'I am delighted to say that your husband will recover.'

That evening however, Pam was asked to go to the doctor's office. This was just 12 hours after my miraculous recovery. The doctor said to her, 'Mrs Dunphy, I am very sorry to tell you that your husband has contracted bacterial pneumonia, and I have arranged for a second opinion from another specialist.' Pam said that she still had absolute faith I would be healed.

The following morning they did a tracheotomy (a small incision in the windpipe) to help me breathe. Coming out of surgery I slipped back into a coma. Pam's cousin knew the second specialist they had called in and asked for his diagnosis. He said that I had a very serious condition which I might not survive.

But Al Ryan had not given up. At six in the evening he came into the hospital holding a big Bible. He and Pam came into the intensive care unit where he prayed for me. He put his hands on me and began praying. Then he stopped. 'Gerard's spirit is gone.' He went on to explain that he was going to pray that if I had not seen the light of heaven, God would allow me to return.

He prayed quietly for a few minutes and then began to pray for healing. As he did, Pam could feel heat coming from my chest, and my eyelids began to flicker just as they had done on the previous Sunday. Then I heard Al say, 'You know who's here, don't you?' I nodded my head and he gave a prayer of thanksgiving. I put my hands together and endeavoured to mouth the words Al prayed.

The following day I was awake and alert. Three days later when the chest specialist returned, he was delighted to see me up and walking about. The tubes in my chest were removed after one week since I was improving. Two weeks later I was released from hospital.

Of course I was aware of very little of what went on during that time, as far as the hospital and surroundings were concerned. What I do know is that during the second coma I had a sensation of my spirit leaving my body. I felt drawn to a light at the end of a long tunnel. To my left I saw a white horse in full gallop. Then I came to a tunnel to the right. I stopped in front of it and heard a voice. It asked me three questions about my love for my wife and children, after which it went on to say that love was the key to the kingdom. It also said I would have to come back to earth. I then returned to my body and regained consciousness.

Out of gratitude, Pam started attending prayer meetings reguarly. At first I went along too, but soon began to feel like an exhibit. After a while I stopped going. They did not give up on me though; they continued to pray for me. Nine months later I asked my wife to get a babysitter so that I could come with her to the prayer meeting that night.

About a year later, we went to a Full Gospel Business Men's Fellowship convention in Blarney. On the Saturday evening I went to the front and asked one of the men to pray for me. I committed my life to Jesus Christ that night.

As that man continued to pray for me, I reached into my pocket, took out my cigarettes, threw them on the floor, and stamped on them. I'd tried unsuccessfully to stop smoking before. Since that time I've not touched a cigarette – the urge to smoke went completely.

Before this experience, my whole life was in a mess. If God had not stepped in, Pam and I would surely have separated. In the period after my miraculous healing and my commitment to Jesus, God began dealing with my problems one by one. From that time I have experienced God's help and guidance in all areas of my life.

19 THE AUSTRALIAN

(The story of Jessie Draper,
now living in Hertfordshire, England)

All my life I had known rejection. I was (according to my mother) an unwanted pregnancy, and I weighed only three pounds when born. I was unhappy throughout my childhood and many times wanted to commit suicide. In fact it was as if there were a death wish on my life.

In 1961 I became a Christian and our first child was born in 1962. I had married Malcolm and we lived on 20,000 acres which we farmed, 105 miles away from any doctor. I had a second child, then four miscarriages before carrying my third child, Wendy.

Despite all my previous problems during pregnancies, Malcolm and I felt very confident about this baby, everything had gone well, and there were no mad dashes over rough roads to the doctor. Of course I knew I must have the baby in hospital and went to our nearest city.

It was then that my problems started. Following the delivery I had a retained placenta and I was haemorrhaging. Naturally, they rushed me to the operating theatre. For two and a half hours they tried, without success, to remove the placenta. At that point they placed the baby girl in my arms and I could feel myself drifting in and out of consciousness.

Suddenly I felt myself being drawn up out of my body, going up and up. Eventually I came to what looked like a sea of glass.

The beauty of the place was astounding. I walked past this sea and I could then see the river of life, just as I'd read about in the Bible. On either side of it there were the most beautiful trees. So overwhelmed with the magnificence of everything surrounding me, I came to a standstill and as I did, I saw Jesus coming towards me. Behind Him, and stretching as far as the eye could see, were thousands of angels, all singing. The music and singing was indescribable. I have never heard such sounds on earth and I wanted to remain standing there, just drinking it in, but I felt myself being drawn further on.

Someone came alongside and put their arms around my shoulders. As He turned me round, Jesus said, 'It is not your time yet, I have work for you to do.' No sooner had He spoken those words than I was back in the operating theatre and I could see myself holding the baby as I came back into my body. The doctors were standing over the other side of the room when all of a sudden something caused them to turn and quickly move over to the bed. They took my blood pressure and this time it was normal.

This experience, amongst many others, caused me to realize that despite the rejection of my childhood, Jesus loved me, and that for Him there is no such thing as an unplanned or unwanted child. God has used me, along with my husband, to bring that love of Jesus to many people, especially young people, in both Australia and England.

20 THE ACTOR'S SON

(The story of Oden Fong, California, USA)

My father was the Hollywood actor, Benson Fong. He starred in the Charlie Chan film series as well as *Keys of the Kingdom* and *Thirty Seconds Over Tokyo*. Because of his lifestyle I was raised around celebrities such as Gregory Peck and Jack Lemmon all the time.

For much of my younger life, my father wasn't around very much, but when he was, he was very strict. I was his only son and so he expected a lot from me. I suppose it was when I was about 13 that I started to rebel. At first it was drinking whisky, then smoking pot regularly, eventually graduating to LSD. As so many addicts do, I became a drug dealer, selling mainly LSD, marijuana, and hashish.

After a few years I moved to Laguna Canyon which was a notorious Orange County drug area, and became close friends with Rosemary Leary, wife of LSD guru Timothy Leary. He was the man who coined the phrase, 'Turn on, tune in, and drop out.' I also delved into Eastern mysticism, using drugs to enhance my 'spiritual' experiences.

I was very serious about what I was into. Once I fasted for 45 days and nights. I did yoga for hours, sitting out on the rocks in the high desert at the Joshua Tree Monument, a national park close to Palm Springs. Sitting on a blanket I would just watch the world turn, trying to 'centre' myself and realize perfection.

There was almost nothing that I did not try in my spiritual quest. I finally got so frustrated that I decided I really wanted once and for all, to become *one* with God. So there at the Joshua Tree Monument in 1970, I took a vial of pure crystal LSD. I snorted the equivalent of 150 doses and immediately began convulsing. I fell over on my back and couldn't breathe. I remember dying, and, according to those with me, they knew I was dead.

During this time I remember my heart beating fast and then ceasing. I peacefully nodded off into death. Then everything started to get dark because I couldn't see any longer. I looked for the light and wanted to head for that light. That was what the mediums and psychics I knew had told me to expect.

When I could see no light, in desperation I cried out. 'Jesus, if you're real, save me.' Then something extraordinary happened. In that darkness, there was light, flashing light that became brighter and brighter. Pretty soon I woke up and could make out the figure of a man in front of me. The brightness was such that I couldn't bear to look at Him. Even the sun, shining behind Him, looked dim by comparison. I spun round and tried to bury my head in the sand but felt as if whatever it was could see right through every part of me. There was no place to hide. I heard His voice saying, 'I am the Alpha and Omega, the beginning and the end.' At that moment in time, I came back into this world.

Soon after this, a group of my old friends who had previously converted to Christianity took me to hear a man preach in Long Beach. I remember listening carefully to what he was saying and knowing that it was what I had to do. I had to give my life totally and completely over to Jesus Christ. At the end of his talk he asked people to come forward for prayer if they wished to make that commitment. I did just that.

The day I 'tuned in, dropped out, and turned on', was the most important day of my life. From that moment on, my life was 'tuned' into the highest force in the universe – Jesus Christ.

21 THE OIL MAN

(The story of David Bremner, United Kingdom)

It was bad timing to say the least. There I was having a massive heart attack and for the first time ever all the local ambulance crews were on strike. And we lived 20 miles from town!

From the early 1950s I had worked in the construction industry and by 1979 when I suffered the heart attack, I was a senior manager in a major oil company, working very long hours, with lots of travelling both on and off shore. I married my childhood sweetheart Jean in 1951. We were God-fearing, church-going people, comfortably off and enjoying good health. One of the requirements of offshore industries, whether working or visiting oil platforms, was to undergo a very strict medical examination at least every three years. I'd had one of these in March of 1979 and whilst a moderate smoker and minimal spirit drinker, I was pronounced as fit as a fiddle.

That year however, the work load became intense and by September I had not had a day off. I was constantly travelling all over the country by plane, train and car and also flying by helicopters to offshore rigs. After a great deal of persuasion from my wife and secretary, I agreed to take a few days off (well, a weekend) and arranged to play golf on the Sunday morning with some friends. I did enjoy a lazy Saturday morning but at lunch time there was a phone call from the offshore construction manager on Brent Bravo about a potential labour dispute.

Because of the direct radio-phone system we had with the rigs, it was easier for me to go into the office to link up and monitor the situation, and issue advice and instructions as and when required.

I did try to unwind that night and the following morning was up about 5.45 a.m. to get ready to play an early round of golf. I decided to make Jean a cup of tea but as I poured out the hot water I had a severe pain in my chest and numbness in my left arm. It eased a moment as I walked through to the bedroom but it then returned with a vengeance, making breathing and moving difficult.

I lay down on top of the bed and felt myself sinking into it. Sounds started to recede as if I were going deaf, and I was only partly conscious. My doctor and friend arrived and I heard him comment to Jean that my hard work had finally caught up with me. He gave me an injection to make me more comfortable and because no ambulance was available, they had to call a police van! Not only were the ambulance men on strike but the unit for treating heart attack victims had been temporarily closed on the Friday as they were opening a new unit on the Monday. I was barely aware of the journey or arriving at the hospital, because what I was experiencing was something very different.

It must have been when my heart stopped that I found myself in a very bright, and white, place and there was no feeling of pain, discomfort or distress; in fact, I knew it was where I wanted to be. I was experiencing a lovely feeling of peace when I became aware of someone saying, 'Come this way David.' In many ways it was like a television set where the scene would change as the channel changed, and I found myself looking down from the top corner of a room, floating just under the ceiling. Beneath me were a group of people gathered around a hospital trolley, talking together very excitedly, but they seemed to be far away. Some were dressed in green, others in outdoor clothes, and one man wore a dress shirt with fancy patterned braces on. On the wall, quite high up, I was conscious of an unusual clock; it was almost opposite me and I thought it rather odd. The man in the fancy braces had something in both hands with wires attached and mumbled something to the others whilst placing these things on the person on the trolley.

Immediately I felt intense pain return and everything went black. Then as light returned I looked down and suddenly realized that it was me on the trolley and I was very angry at what was happening. I wanted to stay where I was because it was peaceful and pain free and I was waiting for someone to come and get me. But once more the man in the fancy braces bent over the trolley and again I felt intense pain, but then nothing until I regained consciousness on the Tuesday.

When I came round there were Jean and Mike, the man whom I had been with on the Saturday sorting out labour problems. He had arranged to come onshore because he felt he had to be at the hospital with me. I started to tell them about the strange experience that I had had but could see by their faces that it was being accepted only with a large pinch of salt. Jean casually remarked, 'That's God giving you another chance. You'll definitely have to change and slow down in future.' Mike, being his usual cynical self, ignored what I was saying and supported what Jean had said about slowing down.

At that point they had to leave because the consultant had arrived – I was now awake and talking. I was also asking for something to eat and drink as well as a wash and shave because I knew I must look awful. Whilst he was checking me over I said to him, 'You have quite a taste in braces.' He stopped and asked, 'What do you mean?' I explained about what had taken place and I could see he was astonished. He wanted to know if someone else had been speaking to me since I had come round but the nurse assured him that she had called him first, as soon as I had opened my eyes.

The consultant explained that it was by chance that he had been in the hospital on Sunday when I was brought in, and that I had died on the trolley and they had done everything to get me back into the land of the living. What I had called jump leads was actually a defibrillator he had used to try and get my heart going again, and yes, he did have on a dress shirt and fancy braces. But how did I know that, as I was not technically alive at the time? As already mentioned, the room I was in was not the one he was used to and when I mentioned the clock, he had not noticed it. Later that evening he returned to say that he had been down to the room concerned and there was a clock just as and where I had described it…

22 THE NURSE

(The story of Pamela Worsey, Bath, England)

In July 1969, already suffering severely with systemic lupus erythematosus (an uncommon disease of the connective tissue), I became acutely ill with a kidney infection and my temperature was creeping up to 105 °F. I began having a kind of nightmare. At the time, I remember thinking I must be delirious with the fever, but if this was so, it was strange that I was lucid enough to consider it! All I can vaguely remember was of horrible, frightening scenes flashing briefly before me.

I then looked up and saw a little window or opening about 30 feet square and I knew I was being given a glimpse into heaven. Through the opening was blue sky and in the left-hand corner was some very beautiful blossom. I suppose the nearest thing I could liken this to would be very wonderful apple blossom. In the top right-hand corner was a dazzlingly brilliant, white light and I knew this was God. He did not speak to me at all during the experience, but I had a one-sided conversation with Him! My first feelings on realizing that I must be either dead or dying were of absolute shock and amazement. At the same time I realized I was exactly the same person a moment after I was dead as I was before, and my first explosive comment of surprise to God was, 'I didn't expect that kidney infection to kill me!'

My second comment was, 'What's going to happen to the faith of all those people down there who I've told You're going to heal me?' Two years previously, following the diagnosis of SLE after years of deteriorating health, major surgery and many, many tests, God had promised me this. When I was praying for healing, God said that if I would 'rest in Him', He would give me my heart's desire (Psalm 37). However, in spite of this and following a trip to a Christian healing centre at Burrswood in Kent, I was going downhill fast, and humanly speaking death looked inevitable. Lying in bed, feeling so ill, I cried out to God, particularly for my husband, and my children who would be left motherless. I'd had a horrendous childhood as I grew up without a mother, and I didn't want the same thing to happen to them. I struggled with God over this, but knew that He is God and all things are possible to Him. I finally surrendered, though I couldn't understand it since I had been promised healing. I told the Lord that though I didn't know how, He would replace me for the family, He could do it, and so if it was His will for me to die, I was willing.

My third comment to God related to the above. I was by now in tears as the full reality of the fact that I was dead or dying hit me and I despaired for my family who would be left bereft. Tearfully, I then said to the Father, 'I really thought I meant it when I said I was willing to die if that was Your will, but now it's come to the crunch, I realize I am not as willing as I thought I was. Please make me willing.' (I remember the conversation word for word!)

The vision then faded and somebody – I presumed it was an angel – took me by the right hand and led me up the hill to Calvary, the place where Jesus was crucified. Unlike so many pictures of Calvary, which look from the front, we approached from the left side, and the whole scene was in darkness, lit eerily by the only light which was coming from behind me from where the angel was. I didn't see the crosses, though I may have had a glimpse of the base of Jesus' cross. I knew where I was and what

I was looking at, and I burst into tears at the pain of it all and pleaded with the person who had led me there, 'Please, don't make me look.'

The scene immediately changed and I was running along the northern side of the chapel at Burrswood. As I ran along a covered, cloister-type way which I do not recall from when my husband took me there to receive ministry, it felt almost as though I was flying – I was so weightless and the movement was so easy. I entered the chapel, presumably through the wall of the north transept, and was immediately alongside people kneeling at the communion rail to receive the laying on of hands for healing, as I had done myself. At once I started to pray for them and then, as I looked up, I saw Jesus! It was just His head and shoulders, with His head turned slightly from me. Gasping inside with the sheer wonder of it all, and unable here to adequately express the awesome joy of this sight, I gasped out to Him, 'Oh Jesus, is it really You?' He turned and smiled at me so lovingly, answering, 'Yes Pam, it's Me.' I was in turmoil as my finite mind tried to grasp the wonderful reality of this experience. I was mentally pinching myself to be sure it was really happening, meanwhile thinking that I must believe this and have faith that it was really happening!

As though Jesus knew exactly what I was thinking (which of course, He always does!) and the inner turmoil I was experiencing, the scene faded and a backcloth of brick red appeared. On this, as I watched, something or someone wrote – in smoke – the letters IHS. They actually appeared as though being written in the normal way with ink, but these were in the same colours of gold, silver and white, as the clothes Jesus had been wearing.

I know some people give the well-known lettering IHS the English rendering of 'In His Service', but Jesus knew that I would know this was a sign from Him to confirm His appearance to me.

The day following this experience I was taken into hospital, but when I came out, I looked up IHS in the Oxford Dictionary. There were several different translations of *In Hoc Signo*, but the one that leaped from the page for me, the only time anything other than God's Word has done this with such an impact and made me sure it was God speaking, was the version, 'In this sign you will conquer'.

23 THE MINISTER

(The story of Pastor Rod Lewis, Birmingham, England)

On February 7 1985, I died in the kitchen at home in front of my 14-year-old daughter.

It had snowed heavily and I had been shovelling the snow to clear a path when I felt a terrific pain like a knife in my back. I shouted out and my wife, Pauline, and my daughter came out to where I was and dragged me into the house. Pauline thought I'd had a stroke and ran to the neighbours for help.

During the time my wife was out of the house, my daughter saw me die. She said that suddenly my body went rigid and my arms and legs just shot out; I stopped breathing and my face changed colour. Not surprisingly, she became hysterical!

I felt myself leaving my body as I went into a tunnel. The sensation was of travelling at high speed down this tunnel and all my natural senses were greatly heightened. The colours were more vivid; my hearing was incredibly acute and everything I saw was manifestly more distinct. I had no fear because, being a Christian, I knew the place to which I was travelling was heaven.

Something I really love to do is to travel fast and so the speed at which I came through the tunnel was a wonderfully exhilarating experience. Approaching the end of the tunnel I saw before me the most brilliant, glorious, radiant light. As I saw the glory of heaven appearing before me, I heard the name of Jesus, seemingly echoing throughout heaven. 'Jesus, Jesus.' On the third shout of 'Jesus,' I did, as it were, an emergency stop and my spirit went into reverse. Back I came up the tunnel at top speed, and re-entered my body.

I became aware of my wife saying, 'Jesus.' Then I saw that she and our neighbours had gathered around me. Later I told them that I knew then I wasn't in heaven because they certainly weren't angels!

It was a most extraordinary experience. I spent some days in hospital and was checked very carefully many times. They said the problem had been caused by a furring of the arteries to the brain. I was due to fly off to the United States to preach a few days later but the specialist told my wife to keep me at home because I was in no state to be going anywhere. However, they took further X-rays and these showed there was nothing stopping the blood flow, so a few days later I flew off to the States.

As I look back over the 10 years since this amazing occurrence, there have been a number of occasions when God has used me to heal other people.

On one such occasion, I was preaching one night in my church at Burrswood and a man died. (That's how boring my preaching is!) About half way through my sermon, he got up and walked out, and I just assumed he needed a break! Having got through the doors, he then collapsed. I sprinted out of the church and there was Jack, laid out, dead. An ambulance had been called but I prayed for him to be healed in the name of Jesus, and instantaneously life came back into his body and he is still fit and well.

I give God all the glory and praise for giving me extra time to serve Him!

24 THE ARMY CADET

(The story of Dr George Ritchie, Texas, USA)

When I was sent to the Army's hospital at Camp Barkeley, Texas, early in December 1943, I had no idea I was seriously ill. I'd just completed basic training, and my only thought was to get on the train at Richmond, Virginia, to enter medical school as part of the Army's doctor-training programme. It was a great break for a private, and I wasn't going to let a cold beat me.

But days passed, and I didn't get any better. It was December 19 before I was moved to the recuperation wing, where a jeep was to pick me up at 4 a.m. the following morning to drive me to the railway station.

A few more hours and I'd make it! Then about 9 p.m. I began to run a fever. I went to the ward boy and begged some aspirin. Despite the aspirin, my head throbbed, and I'd cough into the pillow to smother the sounds. At 3 a.m. I decided to get up and dress.

The next half hour is a blur for me. I remember being too weak to finish dressing and a nurse coming to the room, and then a doctor, and then a bell-clanging ambulance ride to the X-ray building. Could I stand, the captain was asking, long enough to get one picture? I struggled unsteadily to my feet. The whirr of the machine is the last thing I remember.

When I opened my eyes, I was lying in a little room I had never seen before. A tiny light burned in a nearby lamp. For a while I lay there, trying to recall where I was. All of a sudden I sat upright. The train, I'd missed the train!

Now I know that what I am about to describe will sound incredible. I do not understand it any more than I ask you to; all I can do is relate the events of that night as they occurred. I sprang out of bed and looked around the room for my uniform. Not on the bedrail. I stopped, staring. Someone was lying in the bed I had just left.

I stepped closer in the dim light, then drew back. He was dead. The slack jaw, the grey skin were awful. Then I saw the ring. On his left hand was the Phi Gamma Delta fraternity ring I had worn for two years.

I ran into the hall, eager to escape the mystery of that room. Richmond, that was the all-important thing – getting to Richmond. I started down the hall for the outside door. 'Look out!' I shouted to an orderly bearing down on me. He seemed not to hear, and a second later passed the very spot where I stood as though I had not been there.

It was too strange to think about. I reached the door, went through, and found myself in the darkness outside, speeding toward Richmond. Running? Flying? I only knew that the dark earth was slipping past while other thoughts occupied my mind, terrifying and unaccountable ones. The orderly had not seen me. What if the other people at medical school could not see me either?

In utter confusion I stopped by a telephone pole in a town by a large river and put my hand against the guy wire. At least the wire seemed to be there, but my hand could not make contact with it. One thing was clear; in some unimaginable way I had lost my firmness of flesh, the hand that could grip that wire, the body that other people saw.

I was beginning to know too that the body on the bed was mine, unaccountably separated from me, and that my job was to go back and rejoin it as fast as I could. Finding the base and the hospital again was no problem. Indeed I seemed to be back there almost as soon as I thought of it. But where was the little room I had left? So began what must have been one of the strangest searches ever to take place; the search for myself. As I ran from one ward to the next, past room after room of sleeping soldiers, all about my age, I realized how unfamiliar we are with our own faces. Several times I stopped by a sleeping figure that was exactly as I imagined myself. But the fraternity ring was lacking, and I would speed on.

At last I entered a little room with a single dim light. A sheet had been drawn over the figure on the bed, but the arms lay along the blanket. On the left hand was the ring. I tried to draw back the sheet, but I could not seize it. And now that I had found myself, how could one join two people who were so completely separate? And there, standing before this problem, I thought suddenly; '*This is death. This is what we human beings call death, this splitting from one's self*'. It was the first time I had connected death with what had happened to me.

In that most despairing moment, the little room began to fill with light. I say 'light', but there is no word in our language to describe brilliance that intense. I must try to find words, however, because incomprehensible as the experience was to my intellect, it has affected every moment of my life since then. The light that entered that room was Christ: I knew because a thought was put deep within me, '*You are in the presence of the Son of God*'. I could have called Him 'Light', but I could also have said 'Love', for that room was flooded, pierced, illuminated, by the most total compassion I have ever felt. It was a presence so comforting, so joyous and all-satisfying, that I wanted to lose myself forever in the wonder of it.

But something else was present in that room. With the presence of Christ (simultaneously, though I must tell them one by one) also had appeared every episode of my entire life. There they were, every thought and event and conversation, as palpable as a series of pictures. There was no first or last, each one was contemporary, each one asked a single question, '*What did you do with your time on Earth?*'

I looked anxiously among the scenes before me: school, home, scouting and the cross-country track team – a fairly typical boyhood, yet in the light of that Presence it seemed a trivial and irrelevant existence. I searched my mind for good deeds. '*Did you tell anyone about Me?*' came the question. '*I didn't have time to do much,*' I answered. '*I was planning to, then this happened. I'm too young to die!*'

'*No one,*' the thought was inexpressibly gentle, '*is too young to die.*' And now a new wave of light spread through the room already so incredibly bright and suddenly we were in another world. Or rather, I perceived all around us a very different world occupying the same space. I followed Christ through ordinary streets and countryside and everywhere I saw this other existence strangely superimposed on our own familiar world.

It was thronged with people. People with the unhappiest faces I have ever seen. Each grief seemed different. I saw businessmen walking the corridors of the places where they had worked, trying vainly to get someone to listen to them. I saw a mother following a 60-year-old man, her son I guessed, cautioning him, instructing him. He did not seem to be listening. Suddenly I was remembering myself, that very night, caring about nothing but getting to Richmond. Was it the same for all these people; had their hearts and minds been all concerned with earthly things, and now, having lost earth, were they still hopelessly fixed here? I wondered if this was hell. To care most when you are most powerless; this would be hell indeed.

I was permitted to look at two more worlds that night – I cannot say spirit worlds for they were too real, too solid. Both were introduced the same way; a new quality of light, a new openness of vision, and suddenly it was apparent what had been there all along. The second world, like the first, occupied this very surface of the earth, but it was a vastly different realm. Here was no absorption with earthly things, but – for want of a better word to sum it up – with truth.

I saw sculptors and philosophers here, composers and inventors. There were universities and great libraries and scientific laboratories that surpass the wildest inventions of science fiction.

Of the final world I had only a glimpse. Now we no longer seemed to be on earth, but immensely far away, out of relation to it. And there, still at a great distance, I saw a city – but a city, if such a thing is conceivable, constructed out of light. At that time I had not read the book of Revelation, nor, incidentally, anything on the subject of life after death. But here was a city in which the walls, houses, streets, seemed to give off light, while moving among them were beings as blindingly bright as the One who stood beside me. This was only a moment's vision, for the next instant the walls of the little room closed around me, the dazzling light faded, and a strange sleep stole over me ...

To this day, I cannot fully fathom why I was chosen to return to life. All I know is that when I woke up in the hospital bed in that little room, in the familiar world where I'd spent all my life, it was not a homecoming. The cry in my heart that moment has been the cry of my life ever since: Christ, show me Yourself again.

It was weeks before I was well enough to leave the hospital and all the time one thought obsessed me: to get a look at my chart. At last the room was unattended; there it was in medical shorthand: *Pvt. George Ritchie, died December 20 1943, double lobar pneumonia.*

Later, I talked to the doctor who signed the report. He told me there was no doubt in his mind that I had been dead when he examined me, but that nine minutes later the soldier who had been assigned to prepare me for the morgue had come running to give me a shot of adrenaline. The doctor gave me a shot of adrenaline directly into the heart muscle, all the while disbelieving what his own eyes were seeing. My return to life, he told me, without brain damage or other lasting effect, was the most baffling circumstance of his career.

Today over 19 years later, I feel that I know why I had the chance to return to this life. It was to become a physician so that I could learn about man and then serve God. And every time I have been able to serve our God by helping some brokenhearted adult, treating some injured child or counselling some teenager, then deep within I have felt that He was there beside me again.

POSTSCRIPT

Guideposts *magazine, where this account first appeared, has affidavits from both the Army doctor and attending nurse on the case which attest to the fact that Dr Ritchie was pronounced dead on the morning of December 20 1943.*

25 THE ATTEMPTED SUICIDE

(The story of Henrietta, United Kingdom)

I knew it all, as most young people think they do. I wanted to get away from my restrictive past and enjoy life. At first I succeeded. A man – we'll call him Simon – came into my life, swept me off my feet and I became involved in a relationship that at first was wonderful. However, after a time things began to disintegrate and I told Simon I was going to go round the world for six months to get myself sorted out.

So I sold the house, gave up my job and went off intending to go to Australia via the USA. When I arrived in California I stopped off to meet up with some Christian friends I had visited six months earlier and whose Bible study I had attended. Although I had been brought up in a Christian home, I had begun to find the Christian life very difficult, and this in part was because I did not understand the full truth about God's love for me.

We were studying a verse in the book of Hebrews 4:16 which says, 'So let us come boldly to the very throne of God and stay there to receive His mercy and to find grace to help us in our times of need.' Someone had asked what you did when the temptation to sin was beyond you and I perked up and listened because I was always sinning and feeling terrible afterwards. One of the ladies there said we need to cry out to God, even as we know we are sinning, even in the very act of sinning and say, 'Father have mercy on me in this time of need.'

For some reason, during my visit to these friends, I decided I could not go ahead with this six-month trip, I wanted some commitment from Simon. So I rang him in England and said, 'Look it's got to be now or never.' But he told me that basically he wanted out of the relationship. Despite this I bought a ticket and flew straight back to London. By then I was going to pieces and I started thinking about killing myself because I couldn't face life without him. As soon as I got back to London, I rang Simon again and he came to see me. He hadn't changed his mind and although I didn't say so to him, inside I was screaming, 'If you go out of this door now I'm going to kill myself.' He left, not knowing what I was thinking and I said, 'Well that's it. I hate You God, You're against me.'

No one, apart from Simon, knew I was back in London and I had no intention of telling anyone about what I was going to do. I was so angry with God, or at least that's where I put the blame, that immediately Simon left I took a huge overdose. My sister, for some reason, had decided to call round to the flat not knowing I was there, and immediately rang for an ambulance which rushed me to hospital and they managed to save my life.

The next morning when I woke up my sister was sitting by my bed and I thought, 'Oh God, I'm still alive, I'm still here.' My first thought was that I must get out of the place so I could do it again and when the psychiatrist came to see me he must have been able to read my mind. He told me they would not let me out of the hospital as he believed I would only try again, despite my protestations. Meanwhile my sister had been on the telephone to a Christian psychiatrist. He told the hospital that he would keep me in his care and so they released me.

He left me with some very good friends for 24 hours because he had some other work to do. They didn't know what had happened and I convinced them that I was perfectly all right and they gave me a lift back to the flat in which I was staying. At about 11 o'clock that evening, the Christian psychiatrist rang me and said that Simon wanted to speak to me but only when I was among Christians, not on my own. I just thought Simon was a coward and was so angry, I decided again to kill myself. Unfortunately all my pills had been removed from the flat and there was no drink either. I looked out of the window of this Victorian house, but decided not to jump as I might just put myself in a wheelchair rather than do the job properly. Then I thought about a knife but decided it might hurt too much. Eventually I had the 'bright' idea to run a bath and put an electric fire into it.

By now it was about midnight and there was nobody to stop me so I sat down and wrote the conventional suicide note to explain how I felt. Emotionally, it seemed as if I was in a dark tunnel which I had been going down for the past week since I had made the call from California. It was like one of those old railway tunnels which I had gone down only to find it was bricked up – I couldn't go ahead and there was no going back because it was too long, too dark and too far to come out of the tunnel again. It was suffocating me and I was too weak to do anything to save myself; there was just this awful blackness and total despair.

Very slowly I collected my thoughts and decided, because I was a Christian and therefore a child of God, I would pray before I died. I knew that suicide was a sin and I wasn't too sure about whether you still went to heaven because you would actually be sinning at the moment of death. At that moment, I remembered that verse about God helping us in time of need. I went and got my Bible, got down on my knees and opened it at Hebrews 4:16. Instead of screaming on the inside as I had been doing for so long, I prayed that verse and told God, 'Father I come to You in my very hour of need, I come to Your throne of grace and I'm sorry I'm committing suicide Lord, but I can't see a way out.'

No sooner had the words come out of my mouth than I found myself in front of God. People have said I must have died but I'm not sure what happened, all I know is that I was there. I didn't see Jesus, only God and I don't know what He looked like because there was just this warmth and light. Then He spoke and greeted me, and I just said, 'Hi!' His next words may sound funny, but I didn't think so then. 'You're early,' He said to which I responded that I knew I was and I was sorry. God then asked me what I was doing there as it wasn't my time. Such was His love that I poured out my heart and told Him I couldn't go on any longer because I could not stand the pain. 'I know you can't,' He responded, 'but you can't come home yet.'

There was no feeling on my part that I wasn't welcome. I felt that if I walked over to the left side of God's throne, I could have walked into heaven, but again He said, 'You can't come home yet, there are people you need to speak to.'

I turned and saw a group of people in darkness. Actually it was more like two small circles standing in the form of a figure of eight. God showed me that these people were going to hell if I didn't speak to them. There was no argument on my part, as I knew He was telling the truth. I turned round to them again and pleaded with God not to send them to hell, just to send someone else to speak to them instead of me. His reply was, 'I can't, you're going to have to do it.'

How I was going to lead them to God, I really didn't know, all I knew was that it was my job and no one else could do this. Then it hit me just what I was going back to and again I told God that I couldn't go back because I couldn't stand the pain. But there was no choice because heaven is a place of perfection and you can't go in there bearing guilt. As I turned for the last time to leave God's presence He said, 'Ring somebody else.'

The next thing I was back in the sitting room, kneeling on the floor where I had started and the telephone was next to me. I thought of the psychiatrist whose last words had been to ring him up any time of day or night, and I did that but there was no answer – the phone just rang and rang. I threw down the receiver and screamed at God, 'Well, I'm just going to have to die because he's not there.' It was as if I was saying, 'Bad luck God, I'm coming anyway.'

But I felt I had to try and contact Doug, a friend, instead. Three times I started to dial the number and three times I put the phone down half way through. Eventually, I got through the whole number and the words came out, 'Oh Doug, I'm killing myself and I don't want to die.' This was the first time I'd heard myself say those words, and Doug just said, 'I know, Henrietta.' He began to pray for me and as I reluctantly joined in, it was if a ton of bricks were lifted from me and I jumped up and shouted, 'It's gone, it's gone!'

Later I discovered that I was pregnant at the time of the suicide attempt and it dawned on me that had I died then, of course the baby would have died too. The Lord then showed me that the first of the two groups I saw in heaven were the ones to whom I would speak and the group further away from me was the group to whom my child would speak. Although the situation had seemed such a disaster it was wonderful to realize that this child was planned by the Lord and He had people for my child to speak to about His love. My daughter is now just turned 11 and she has brought so many people to know Jesus, it is amazing. Every time she tells me about someone else she has spoken to, I jump for joy because I remember that group of people in heaven.

26 HEAVEN AND HELL – YOUR CHOICE

From the authors

One of the great things about being human is that we have the ability to make choices. Today's society is concerned about the liberty of individuals to decide for themselves. Everyone who has told their story in this book has made a crucial decision in their lives. For some, they had made their choice before their experience of death, others after their encounter with heaven or hell. We've added no comments to the stories, as they speak for themselves, but as Christians we believe it is vital people do think about where they will spend eternity.

We believe that God created us as explained in the first book of the Bible, Genesis. 'So God created people in his own image; God patterned them after himself; male and female he created them' (1:27). We then learn, 'And the Lord God formed a man's body from the dust of the ground and breathed into it the breath of life. And the man became a living person' (2:7).

It is interesting to note that the chemical components of the human body are exactly the same chemical elements as the dust of the ground. Our physical bodies are made up of about 17 chemical elements – the same 17 that are found in the dust of the ground. It is a matter of common observation that, after a body has been cremated, only dust remains.

It appears from reading these first two chapters of the Bible that the creation of the human body was a two part process.

First of all the spirits of Adam and Eve were created in the image of God, and then these spirits were breathed by God into a human body, formed of the dust of the ground. It follows therefore, that we too are flesh, made of the same chemical elements as the dust of the earth, but containing an eternal spirit. At death the human body decomposes, but the spirit which is indestructible lives on.

This means that the real you is a spirit that will live forever. When the physical body dies, our spirit will live on and go either to heaven or to hell. Both of these places have been described quite graphically in many of the stories and no doubt you have already decided that hell is not where you want to finish up!

You might ask what evidence there is to support the ideas of heaven and hell contained in these accounts. There are many verses in the Bible which tell us about both. John 14:2 tells us that heaven is where God lives. In Luke 15:7 we read that heaven is happy when someone turns to God and asks forgiveness for their sins. In the book of Revelation there are many verses which describe how wonderful heaven is, especially in chapters 21 and 22. These describe a new world where there is no crying, no sadness, no pain. Here the streets are pure gold, as clear as glass and the walls are made of beautiful gems.

By contrast hell is described as a place of eternal torment and pain where the body suffers. This is shown most clearly in the story that Jesus told in Luke 16:19–31. A poor beggar, named Lazarus, died and was taken by the angels to be with Abraham (one of the first recorded people in the Bible). In his lifetime, Lazarus used to beg outside the house of a very rich man but he was given nothing by the man of the house. This rich man also died and was buried, and his soul went to the place of the dead. There he was in torment, desperate for water and begged Abraham to have pity on him. 'Send Lazarus over here to dip the tip of his finger in water and cool my tongue, because I am in anguish in these flames.' But Abraham told him that during his time on earth he had had everything and given Lazarus nothing. Between himself and Lazarus, the rich man was told, was a great chasm and there was no way that anyone could cross over it.

This story shows most clearly that once we have died, the decision taken on this earth regarding our eternal destination, heaven or hell, is final.

You may never have the opportunity, as the people in this book have, to glimpse what lies beyond the final frontier of death. But you have read about their experiences. We cannot prove or disprove what they have seen. We can say however, that what they have seen is confirmed by what Jesus told us while He was on earth, and what God revealed to other writers in the Bible.

There is only one way to make certain that heaven is our destination when we die. We have to follow the ways and teachings of Jesus Christ. God loved us so much that He sent His only Son Jesus, the Bible tells us in John 3:16, to die on the cross. If we believe this, accept God's forgiveness, and decide to follow Jesus, then we are assured of eternal life in heaven. If we reject the forgiveness that God offers us then we are turning away from Him and we remain 'sinners'. The Bible tells us very clearly in that same verse that eternal separation from God will be the result of this action.

Romans 3:23 says that we are all sinners in God's sight because we have broken His commandments. God is holy and pure and sin cannot exist in His presence. The only way to God is to say we are sorry for the wrong things in our life and accept the forgiveness offered to us. Jesus promises that He will accept us just as we are.

Listed below are six simple steps that you can take to become a Christian and be certain of eternal life.

1) First, admit that you have lived selfishly and in not honouring God, you are a sinner and separated from Him. (Romans 3:23)

2) Say you are sorry and ask Him to forgive you for all those things in your life that you know are wrong. (Luke 13:3)

3) Tell God that you believe Jesus died on the cross in order to take away your sins, and that you want Him to come and guide your actions and your life. (John 3:16)

4) Tell other people that you believe Jesus is the Son of God and that not only did He die for you, but that God resurrected Him from the dead. (Romans 10:9)

5) Pray simply: 'Dear God, I know that I have sinned and need your forgiveness. I believe that Jesus, Your Son, died for all sinners, including me. Please forgive me as I forgive all those other people who have done wrong to me. I ask You to become the Lord of my life. Thank You for the gift of eternal life. Help me to live a life pleasing to You.'

6) You now have a new start. Don't depend on feelings as proof of your acceptance by God. Feelings are changeable, but your relationship with God is based on His promises. Take time for daily prayer and Bible reading. (Romans 10:13, Matthew 10:32, 1 Peter 2:2, Psalms 37:4 and Romans 8:14)

From the authors

It is important for you to have further information about the Christian life. In this book, mention has been made of the Full Gospel Business Men's Fellowship International, which is an organization of Bible-believing Christians. Please contact them at one of the following addresses and they will send you a booklet entitled 'Now You've Received Christ'.

Full Gospel Business Men's Fellowship International
UK Office
PO Box 11
KNUTSFORD, Cheshire, WA16 6QP
Tel: 01565 632667
Fax: 01565 755639

USA Headquarters
20 Corporate Park, 3rd Floor,
IRVINE, California 92714
Tel: (714) 260 0700
Fax: (714) 260 0718
Internet: http://www.nettap.com/fgbmfi
e-mail: fgbmfi@ix.netcom.com

European Office
Mechelsesteenweg 30
3000 LEUVEN
Belgium
Tel: (32) 16 207944
Fax: (32) 16 207931
e-mail: 100447.1425@compuserve.com

You may also find it helpful to visit your local Christian book shop where you will be helped with suggestions for other books for you to read.

BIBLIOGRAPHY

Ankerberg, J. & Weldon, John, *The Facts on Life After Death*, Oregon, Harvest House, 1992

Baker, H. A., *Visions Beyond the Veil*, Springdale, Whitaker House, 1973

Baxter, Mary K., *A Divine Revelation of Hell*, Springdale, Whitaker House, 1993

Blanchard, John, *Ultimate Questions*, Darlington, Evangelical Press, 1991

Buchanan, Alex, *Heaven and Hell*, Tonbridge, Sovereign Word, 1995

Darnall, Jean, *Heaven Here I Come*, London, Lakeland, 1974

Eby, Dr Richard, *Caught Up Into Paradise*, New Jersey, Spire Books, 1971

Fernando, Ajith, *Crucial Questions About Hell*, Eastbourne, Kingsway, 1993

Graham, Jim, *Dying to Live*, Basingstoke, Marshalls Paperbacks, 1984

Jeffrey, Grant, J., *Heaven The Last Frontier*, Toronto, Frontier Research, 1990

Liardon, Roberts, *I Saw Heaven*, Oklahoma, Embassy, 1983

Lindsay, Gordon, *Death and Hereafter*, Dallas, Christ for the Nations, 1986

Malz, Betty, *My Glimpse of Eternity*, London, Hodder & Stoughton, 1990

Morgan, Dr Alison, *What Happens When We Die,* Eastbourne, Kingsway, 1995

Osteen, John, *Death and Beyond,* Houston, Lakewood, 1985

Pawson, David, *The Road to Hell,* London, Hodder & Stoughton, 1993

Pawson, David, *Resurrection,* Tonbridge, Sovereign Word, 1993

Rawlings, Dr Maurice, *Beyond Death's Door,* London, Sheldon Press, 1979

Rawlings, Dr Maurice, *Before Death Comes,* London, Sheldon Press, 1980

Rawlings, Dr Maurice, *To Hell and Back,* London, Thomas Nelson, 1993

Ritchie, Dr George, *Return from Tomorrow,* Eastbourne, Kingsway, 1992

Torrey, R. A., *Get Ready For Forever*, Springdale, Whitaker House (no date)

Wagner, Dr Petti, *Murdered Heiress...Living Witness,* Chichester, New Wine Press, 1988